GENERALIZED LINEAR MODELS FOR BOUNDED AND LIMITED QUANTITATIVE VARIABLES

Quantitative Applications in the Social Sciences

A SAGE PUBLICATIONS SERIES

Quantitative Applications in the Social Sciences

A SAGE PUBLICATIONS SERIES

GENERALIZED LINEAR MODELS FOR BOUNDED AND LIMITED QUANTITATIVE VARIABLES

Michael Smithson

Australian National University

Yiyun Shou

Australian National University

Los Angeles | London | New Delhi
Singapore | Washington DC | Melbourne

Los Angeles | London | New Delhi
Singapore | Washington DC | Melbourne

FOR INFORMATION:

SAGE Publications, Inc.

2455 Teller Road

Thousand Oaks, California 91320

E-mail: order@sagepub.com

SAGE Publications Ltd.

1 Oliver's Yard

55 City Road

London EC1Y 1SP

United Kingdom

SAGE Publications India Pvt. Ltd.

B 1/I 1 Mohan Cooperative Industrial Area

Mathura Road, New Delhi 110 044

India

SAGE Publications Asia-Pacific Pte. Ltd.

18 Cross Street #10-10/11/12

China Square Central

Singapore 048423

Printed in the United States of America

ISBN: 9781544334530

Acquisitions Editor: Helen Salmon

Editorial Assistant: Megan O'Heffernan

Content Development Editor: Chelsea Neve

Production Editor: Jyothi Sriram

Copy Editor: Gillian Dickens

Typesetter: Integra

Proofreader: Barbara Coster

Indexer: Laurie Andriot

Cover Designer: Anupama Krishnan

Marketing Manager: Shari Countryman

MIX
Paper from
responsible sources
FSC® C008955

CONTENTS

SERIES EDITOR'S INTRODUCTION

Truth be told, many of us have not given much consideration to the kinds of models presented in this volume, but we should. Many of the variables we analyze are bounded. For example, years of education is bounded at 0, percentages at 0 and 100, and time use between 0 and 24 hours in a day. These are "absolute" bounds. Bounds may also be due to truncation and censorship. A measure of social and economic conservatism that asks respondents to "please indicate the extent to which you feel positive or negative towards" a set of issues, with anchors denoting each end of a 0-100 scale as "completely negative" or "completely positive" is an example from this monograph. There is nothing absolute about 0 and 100 in this scale. When it comes to attitudes towards gun ownership, there is a pileup of responses at each end of the scale.

Generalized Linear Models for Bounded and Limited Quantitative Variables acquaints non-specialists with approaches for modeling bounded variables. The focus is variables having continuous ranges with one or two bounds due to absolute limits, truncation, or censoring. Readers have the opportunity to learn from experts in this area. The authors, Michael Smithson and Yiyun Zhou, have been central to the development of these methods, especially for modeling doubly-bounded variables.

Generalized Linear Models for Bounded and Limited Quantitative Variables is a "second course" on generalized linear models (GLMs) with clear explanations and lots of helpful advice. The volume grounds readers in the analysis of bounded variables, equips them with the understanding and tools needed to model these variables, and introduces readers to recent advancements including the authors' own work on CDF-quantile GLMs. Although the authors provide a brief review of GLMs in the first chapter, readers would do well to be already familiar with this material. Those needing a tutorial can find one in the just published second edition of *Generalized Linear Models*, by Jeff Gill and Michelle Torres.

A strength of the volume's pedagogy is an extensive use of examples based on real data from diverse disciplines. Approaches to singly bounded variables are illustrated with analyses of average days to start a business in various countries of the world, individual response times on a Stroop test, and ambulance arrival times in regions of Wales.

Examples of doubly bounded variables come from judgments of the grammaticality of sentences generated by machine translation, probability judgments in a moral dilemma, family factors and depressive symptoms among university students in China, party affiliation and political and social conservatism, and ratings of features essential to democracy. The data come from administrative records, experiments, surveys, and other sources such as mTurk. There is something for everyone. The data for the examples, along with detailed explanations and software code in R, SAS, and Stata, are in a supplementary website so that readers can reproduce the results reported and discussed in the volume.

With this book as a foundation, analysts now have every reason to incorporate the boundedness of variables in their analyses. Its timing is propitious. Software resources for analyzing bounded variables, especially doubly-bounded variables, are increasingly available in a variety of packages, including R, SAS, and Stata. This volume provides the guidance needed to design these analyses and interpret the results.

—*Barbara Entwisle*
Series Editor

ABOUT THE AUTHORS

Michael Smithson is a Professor in the Research School of Psychology at The Australian National University in Canberra and received his PhD from the University of Oregon. He is the author of *Confidence Intervals* (2003), *Statistics With Confidence* (2000), *Ignorance and Uncertainty* (1989), and *Fuzzy Set Analysis for the Behavioral and Social Sciences* (1987); coauthor of *Fuzzy Set Theory: Applications in the Social Sciences* (2006) and *Generalized Linear Models for Categorical and Limited Dependent Variables* (2014); and coeditor of *Uncertainty and Risk: Multidisciplinary Perspectives* (2008) and *Resolving Social Dilemmas: Dynamic, Structural, and Intergroup Aspects* (1999). His other publications include more than 170 refereed journal articles and book chapters. His primary research interests are in judgment and decision making under ignorance and uncertainty, statistical methods for the social sciences, and applications of fuzzy set theory to the social sciences.

Yiyun Shou is a research fellow in the Research School of Psychology at The Australian National University. She received her PhD degree in psychology in 2015 and was recently awarded an Australian Research Council Discovery Early Career Award (2018–2021). She is active in research in the areas of understanding measurement issues in psychology and developing new quantitative methods. She also conducts extensive research in judgment and decision making under uncertainty and cross-cultural psychological assessments. She has publications in a number of respected international outlets for measurement and quantitative psychology such as the *Journal of Statistical Software*, *British Journal of Mathematical and Statistical Psychology*, *Psychometrika*, and *Psychological Assessment*.

ACKNOWLEDGMENTS

The authors would like to thank the following reviewers for their feedback:

Paul E. Johnson, *University of Kansas*

Jeff Harden, *University of Notre Dame*

Stephen Broomell, *Carnegie Mellon University*

Jerald R. Herting, *University of Washington*

William Berry, *Florida State University*

Scott Long, *Indiana University*

Richard Williams, *University of Notre Dame*

James Endersby, *University of Missouri*

Irwin D. Waldman, *Emory University*

COMPANION WEBSITE FOR THIS BOOK

The data, software code for R, Stata®, and SAS®, and detailed explanations of the example models outlined in this book are available on an accompanying website at **www.study.sagepub.com/researchmethods/qass/smithson-and-shou-generalized-linear-models**.

Note:

- Stata is a registered trademark of StataCorp LLC.

- SAS and all other SAS Institute Inc. product or service names are registered trademarks or trademarks of SAS Institute Inc. in the USA and other countries.

CHAPTER 1. INTRODUCTION AND OVERVIEW

1.1 Overview of This Book

This book provides a course on generalized linear models for bounded variables. We focus on numeric dependent variables whose scales are bounded either at one end or both ends. Examples are income (typically bounded below at 0), hours spent on an activity per day (bounded between 0 and 24), or percentage of a population eligible to vote (bounded from 0 to 100).

Why is this topic important? The human sciences deal in many variables that are bounded. Ignoring bounds can result in misestimation and improper statistical inference. On the other hand, taking bounds into account not only can provide more accurate statistics but also often reveals insights that otherwise would escape the researcher's notice.

Why is a book like this needed? Bounded scales and variables often are analyzed with conventional but inappropriate techniques. This is due to lack of familiarity with the techniques for working with bounded variables. Books and other teaching materials that introduce generalized linear models ignore or give only scant attention to such techniques. Instead, techniques and relevant concepts regarding bounded variables typically crop up only in specialized contexts such as survival analysis, psychophysics, engineering, or econometrics. Our goal is to make them generally available and accessible to nonspecialists.

There are book-length surveys of limited dependent variables, including the now classic monograph by Long (1997b) and the more recent treatment by Smithson and Merkle (2013). Books also have been devoted to subsets of these variables. Categorical variables understandably have received the most attention, including handbooks on logistic regression (Hosmer Jr., Lemeshow, & Sturdivant, 2013), log-linear analysis (Agresti, 2013), and ordinal categorical regression (Agresti, 2010). Singly bounded variables usually appear under the guise of "life distributions" (Marshall & Olkin, 2007). Censoring and truncation have been given an accessible introduction by Breen (1996). To our awareness, however, there is no introduction to quantitative bounded variables that is both accessible and as thorough as might be desired. By focusing exclusively on these variables (rather than including them along with categorical variables), we have attempted such an introduction in this book.

This is a good time for this book to appear, because the early part of the 21st century has seen rapid growth in the availability of software resources for analyzing bounded variables. Bounded-variable models are available in popular computing environments, including R, Stata, MPlus, and SAS. Generalized linear models (GLMs) for singly bounded variables have been widely available for some time in these environments. GLMs for doubly bounded variables have been made available in R, Stata, and SAS. The stage is set for accelerating understanding and application of these models, and this book is intended to provide a guide and impetus for such applications.

This is a "second" course in generalized linear models. We assume that the reader is familiar with linear multiple regression and has had at least an introduction to the general linear model. The Sage collection has excellent books on both topics: Colin and Michael Lewis-Beck's (2015) *Applied Regression* and Jeff Gill's (2000) *Generalized Linear Models*.

Here is what our book covers. In this first chapter, we introduce the different types of bounds and present examples of two of the most well-known kinds. In Chapter 2, we focus on singly bounded variables, beginning with the basic concepts underpinning a GLM for such variables and then introducing readers to three of the most popular relevant distributions and examples of models that use them. In Chapter 2, we also discuss alternative methods for dealing with cases on the boundary and how to choose such a method. In Chapters 3 and 4, we cover methods for dealing with doubly bounded variables. Chapter 3 focuses on models employing the beta distribution, while Chapter 4 introduces distributions for modeling quantiles of doubly bounded variables. In Chapter 5, we deal with what are known as "censored" and "truncated" variables, starting with the popular Tobit model and then moving on to different types of censoring and truncation, as well as non-Gaussian and heteroscedastic models. In Chapter 6, we conclude the book with an overview of bounded variables and brief discussions of extensions to the models covered in the preceding chapters. These extensions include multivariate and random-effects models, Bayesian estimation, and the treatment of bounded covariates or independent variables.

We provide worked examples in every chapter, using real data sets from a variety of disciplines. The software used for the examples include R, SAS, and Stata. The data, software code, and detailed explanations of the example models are available in the supplementary materials on the website for this book: www.study.sagepub.com/researchmethods/qass/smithson-and-shou-generalized-linear-models. We take care not to include particulars in the book about software that quickly become out-dated, instead relegating such specifics to the supplementary web-based materials where they can be updated as needed. The supplementary materials also contain additional details, including code, development, and interpretation of the models, in some cases going beyond the models described in the book itself.

1.2 The Nature of Bounds on Variables

Bounds on variables occur in two contexts: as categorical bounds and bounds on one or more continuous ranges. This book deals with variables that have bounds on continuous ranges. Moreover, we restrict our treatment to variables that have only one or two bounds. Bounded variables also require two kinds of considerations regarding GLMs for them. First, there is the problem of constructing a GLM that takes the bounds into account (e.g., by not generating predictions outside of the range). Second, there is the issue of how to model and interpret cases at the bounds.

A useful typology of bounds on continuous variables distinguishes among three kinds: "absolute," "censoring," and "truncating." Absolute bounds are values beyond which it is impossible for the variable to go (e.g., a proportion cannot go below 0 or above 1). Censoring bounds are values that only put a lower or an upper limit on the true scores of cases on the boundary (i.e., those cases' scores are "censored"). For example, a webpage automatically logs visitors out after their inactivity has exceeded 15 minutes, so lengths of visits to that webpage are only known to be at least 15 minutes' duration if they timed out. Truncating bounds are those for which cases are excluded altogether from a sample (e.g., a perception experiment excludes participants whose vision is worse than 20-40). Absolute bounds more strongly constrain the choice of distributions for modeling the data because they determine the support or domain of the distribution, whereas censoring and truncation do not entirely determine the distribution support.

Among the most common kinds of censoring or truncating are when bounds on a variable are an artifact of a nonexhaustive collection of items whose scores are combined to form a scale or due to scale endpoints that truncate scores. An example of the first kind is the ethical risk-taking subscale in the DOSPERT (the Domain-Specific Risk-Taking Inventory; Weber, Blais, & Betz, 2002), composed of eight items, each of which has a score from 1 to 5, so that the subscale range is from 8 to 40. We cannot infer that a person who scores 8 on this scale never takes any ethical risks, because the eight items do not exhaust the list of unethical acts. An example of the second is a scale recording household incomes with an upper bound "more than I," where I is a threshold amount.

There also are different varieties of truncation due to sample selection. Boundary cases may consist exclusively of exclusions from a sample, such as amounts owing on household mortgages in which the zeros are those households that are not currently mortgaged. Conversely, boundary cases may include a mix of sample exclusions and inclusions. For instance, data consisting of the number of cigarettes smoked by a person in the past week may include zeros that are smokers as well as zeros that are nonsmokers.

The question of whether bounds are absolute or not can be debatable for at least three reasons. First, the answer may depend on the target construct being measured by the scale. Percentage score on an examination is a commonplace example. Clearly, it makes no sense for a percentage to fall below 0 or above 100 in this context, and if the exam is intended simply to measure knowledge about what is being examined, then these bounds are absolute. However, if the examination items do not exhaust the subject of the examination, then the bounds are not absolute in the sense of measuring the student's knowledge of the subject. A student scoring 0 still may have some subject knowledge, and a student scoring 100 may not know everything about the subject.

Second, the definition of the construct may determine the nature of the bounds. In Tobin's (1958) original application of a censored regression model, the dependent variable was the ratio of household expenditure on durable goods to disposable income. If we regard the underlying construct as propensity to purchase durable goods, then we may elect to define an observed value of 0 as corresponding to a latent propensity of 0 (so the boundary is absolute) or to define it as a censored latent propensity.

Third, the boundaries may be open to multiple interpretations by respondents. An example is a scale with verbal anchors, such as a World Values Survey (World Values Survey Association, 2015) item that asks participants, "How important is it for you to live in a country that is governed democratically?" on a scale from 1 to 10, where 1 is labeled *not at all important* and 10 is labeled *absolutely important*. Taken literally, this scale's bounds seem absolute insofar as values outside these bounds do not make sense. However, there is no guarantee that all respondents will interpret these verbal anchors literally.

Finally, we return briefly to the issue of boundary cases. A preponderance of cases on a boundary produces what is sometimes called a "boundary-inflated" distribution. It suggests that at least some of the boundary cases may be distinct in some way from the rest of the cases. In some contexts, there are good reasons for supposing the existence of such a distinction. Suppose we ask people to estimate the probability that humanity will become extinct within the next thousand years. A large number of zeros in the responses would indicate that there may be two types of respondents, those who believe the human species never will become extinct and those who believe that extinction is possible. We may then decide to ascertain what distinguishes the zero-respondents from the others, separately from identifying predictors of how probable people think human extinction is. Another important class of boundary cases is so-called corner solutions, as identified in the economics literature. A corner solution occurs when an agent maximizes his or her utility at a boundary, as in a person who refuses all bets, no matter how attractive, on grounds that it is against his or her moral code to gamble. We will revisit the treatment of boundary cases several times in this book. First, however, we will briefly review the concepts from the generalized linear model that will be used throughout this book.

1.3 The Generalized Linear Model

1.3.1 Definitions and Concepts

Many of the models described in this book are generalized linear models, or GLMs (McCullagh & Nelder, 1989). We are going to provide a brief introduction to them, starting with definitions and basic concepts. For a complete introduction to GLMs, however, readers should consult Jeff Gill's book (2000). First, let us consider the *general linear model*. Suppose we have a dependent variable, Y, whose expected value is a linear function of predictor variables x_j, for $j = 1, \ldots, J$:

$$Y = \mu + e = \beta_0 x_0 + \beta_1 x_1 + \beta_2 x_2 + \cdots + \beta_J x_J + e = x\beta + \epsilon, \quad (1.1)$$

where the vector x has $x_0 = 1$ as its first entry, β is the vector of coefficients, and ϵ has a normal distribution with mean 0 and variance σ^2. Another way to write the model in equation (1.1) is to think of Y as having a *conditional distribution* (i.e., a distribution whose parameters are at least partly determined by the predictor variables). So, in equation (1.2), we describe the conditional distribution of $Y|x, \beta$ as normal with mean μ and variance σ^2, where μ is determined by the weighted linear combination of predictors, $x\beta$. This last equation in equation (1.2) is the *systematic component* of the general linear model, and the ϵ error term in equation (1.1) is what makes $Y|x, \beta$ the *stochastic component*.

$$Y|x, \beta \sim N\left(\mu, \sigma^2\right)$$
$$\mu = x\beta. \tag{1.2}$$

The *generalized linear model* often is confused with the general linear model, but it is indeed a more general form of a linear model in two respects. First, it relaxes some of the assumptions required of the general linear model and admits other distributions than the normal. Second, it involves a *link function* connecting the systematic and stochastic components of the general linear model. The link function, g, is applied to the parameter μ being estimated via the weighted linear combination of predictors, as shown in equation (1.3). This function is smooth and monotonic in μ. In some representations of the GLM, the inverse is used instead (i.e., $\mu = g^{-1}(x\beta)$) because it focuses on the estimation of the parameter μ. The general linear model, then, is a special case of the GLM, with the link function being the identity: $g(\mu) = \mu$.

$$Y|x, \beta \sim f\left(\mu, \sigma^2\right)$$
$$g(\mu) = x\beta. \tag{1.3}$$

In some cases, there may be more than one available link function. A special link function is known as the *canonical link*, because it arises as a result of how the distribution f is defined. Why is this distinction between canonical and noncanonical link functions important? Gill (2000) provides a detailed and accessible explanation of the desirable statistical properties of the canonical link. Aside from the statistical or theoretical reasons, the most pragmatic reason is that in a few cases, the domain of the canonical link function is not the same as the permissible range of μ. For example, this is true of the exponential and gamma distributions, whose canonical link functions are the reciprocal functions. In such cases, modelers often will use an appropriate noncanonical link function (such as the log) that restricts estimates to the permissible range.

As an example, we introduce the lognormal GLM, where the log of Y is assumed to have a normal distribution. We are using this example because the lognormal is a popular choice for modeling variables that have a lower bound of 0, such as income or reaction time. In equation (1.4), we see that Y is distributed as lognormal (LN) with parameters

μ and σ, and the link function $g(\mu) = \log(\mu)$. Although μ can only take positive values, $\log(\mu)$ can take any value on the real line, positive or negative. Thus, g unbounds the μ parameter so that it can be estimated by the weighted linear combination of predictors without worrying about the lower bound.

$$Y|x, \beta \sim LN\left(\mu, \sigma^2\right)$$
$$\log\left(\mu\right) = x\beta. \tag{1.4}$$

It is important to bear in mind that the link function does not always translate straightforwardly to the expected value of Y or its variance. In the lognormal distribution, although $E(\log(Y)) = \mu$, the expectation of Y is $E(Y) = \exp(\mu + \sigma^2/2)$. Likewise, although the variance of $\log(Y)$ is σ^2, the variance of Y is $(\exp(\sigma^2) - 1)\exp(2\mu + \sigma^2)$. This type of difference between the link function and the parameterization of summary statistics such as the mean, variance, and quantiles will crop up frequently for bounded variables, and we will return to this issue several times to extract the connection between it and the nature of bounds on variables.

Most of the GLMs in this book involve *location-scale* distributions that are fully specified by two parameters. Often, these consist of a location parameter (e.g., a mean) and a scale or dispersion parameter (e.g., a standard deviation). Roughly speaking, a location parameter determines the central tendency of a distribution, whereas a dispersion parameter determines the variability of a distribution.

Traditionally, applied modelers have focused exclusively on modeling the location parameter, relegating the dispersion parameter to the status of a nuisance parameter or a role in evaluating location model error. This tradition stems from two related sources. One is the popular assumption of homoscedasticity or homogeneity of variance (i.e., variance is constant regardless of the location of the mean), as in conventional linear regression models.

The second, more implicit source is the fact that many location-scale distributions whose support is the real line (e.g., the normal and t distributions) have the property that the location parameter can be changed without altering dispersion and vice versa. However, for variables with bounds, this generally is not the case. When location and dispersion are not independent of one another, then modeling both location and dispersion parameters becomes important.

Returning to the lognormal GLM example, we already have seen that in its original scale, the mean and variance of Y are influenced both by μ and by σ. Thus, in its original scale, Y is naturally heteroscedastic.

It therefore could make sense to model both μ and σ explicitly, with predictor variables for each. We then would have two *submodels*, as shown in equation (1.5), a *location submodel* for μ and a *dispersion submodel* for σ, each with its own link function (in this example, it is the log for both). Moreover, in $z\delta$, the variables in the vector z may differ from or overlap with those in x.

$$
\begin{aligned}
Y|x,\beta &\sim LN\left(\mu,\sigma^2\right) \\
\log(\mu) &= x\beta \\
\log(\sigma) &= z\delta.
\end{aligned}
\tag{1.5}
$$

1.3.2 Estimation

Most of the GLMs in this book are estimated via maximum likelihood (ML) estimation rather than least squares estimation. The latter has been the default method in linear regression, and in linear regression, least squares and ML estimates are identical. However, in many GLMs, especially those that model dispersion parameters, ML estimation is relatively straightforward whereas least squares estimation often is inapplicable. Equation (1.1) presents a "least squares" view of a GLM with the conventional error term ϵ, whose sum of squares is minimized in least squares estimation. On the other hand, equations (1.2) to (1.5) present a "likelihood" view that refers directly to a distribution, conditional on its parameter values. The error term ϵ has been absorbed into the conditional distribution, and by the time we arrive at equation (1.5) with its submodel for σ, there is no viable place for an error term of that kind.

Instead, the ML approach deals with the likelihood of each observation in the distribution, conditional on the parameter estimates. Suppose we have a random sample of independent observations x_i, for $i = 1, \ldots, N$, from a random variable X whose probability density function is $f(x,\theta)$, where θ is the vector of parameters that define the density function. Then the likelihood of any x_i conditional on μ and σ is $f(x_i|\theta)$. The *likelihood function* is the product of all of these likelihoods, due to the independence of the x_i. The log of the likelihood function is more computationally convenient to work with, so the ML approach conventionally uses the log-likelihood function

$$
L(x|\theta) = \sum_{i=1}^{N} \log(f(x_i|\theta)).
\tag{1.6}
$$

The *maximum likelihood estimates* of the parameters in a GLM are the values of those parameters that maximize $L(x|\theta)$.

1.3.3   Evaluating and Comparing Models

Given alternative models with different sets of parameters and predictors, researchers frequently wish to compare these models to determine which of them fit the data best. This inclination naturally arises from the availability of likelihood statistics and some of their attractive properties, and while it may not always result in the best practice regarding comparisons among models, we will use it throughout this book because it is practical and also related to Bayesian model comparison methods such as Bayes factors.

To begin, it is important to distinguish between *nested* and *nonnested* models. Model 1 is nested in Model 2 if Model 2 includes all of Model 1's parameters and additional parameters as well. For instance, suppose that Model 1 has a location submodel $\mu = \beta_0 + \beta_1 x_1 + \beta_2 x_2$, while Model 2's location submodel is $\mu = \beta_0 + \beta_1 x_1 + \beta_2 x_2 + \beta_2 x_3$ and Model 3's location submodel is $\mu = \beta_0 + \beta_2 x_2 + \beta_2 x_3$. Then Model 1 is nested in Model 2 but not in Model 3, whereas both Models 1 and 3 are nested in Model 2. Sometimes the nested model (e.g., Model 1) is called the *reduced model*, and the nesting model (Model 2) is called the *full model*.

When models are nested, they may be compared via likelihood-based tests. There are three asymptotically equivalent such tests: the likelihood ratio, Lagrange multiplier, and Wald tests (Engle, 1984). The likelihood ratio test (LRT) may be written as

$$G_{12}^2 = -2\left(L\left(y\,|x_1,\theta_1\right) - L\left(y\,|x_2,\theta_2\right)\right), \tag{1.7}$$

where the subscripts 1 and 2 denote the parameters in Models 1 and 2, respectively, and all of the variables and parameters in Model 1 are contained in those for Model 2. When the null hypothesis that these model likelihoods are equal (i.e., the additional parameters in Model 2 do not improve model fit) is true, G_{12}^2 asymptotically follows a χ^2 distribution, whose degrees of freedom are $k_2 - k_1$, where k_1 and k_2 are the number of parameters in Models 1 and 2, respectively. Rejecting the null hypothesis motivates the researcher to regard Model 2 as better fitting than Model 1. Failing to reject the null hypothesis, on the other hand, justifies preferring Model 1 over Model 2 on grounds of parsimony.

The other two tests sometimes are useful alternatives to the LRT, partly because each of them requires full estimation of only one of the alternative models. The Lagrange multiplier test requires only that the reduced model be estimated. The full model is then fitted in a single iteration, and the resulting change in fit is used to evaluate the full model.

The Wald test, on the other hand, requires only complete estimation of the full model. It simultaneously tests null hypotheses regarding parameters in the model, enabling the researcher to discard those parameters for which the null hypothesis cannot be rejected. However, the Wald test uses two approximations (the χ^2 approximation and approximate standard errors for the parameters), whereas the LRT uses only the χ^2 approximation, so the Wald approach is more susceptible to small-sample inaccuracy.

These three tests cannot be applied to comparisons between nonnested models. The two most popular statistics for comparing nonnested models are the *Akaike information criterion* (AIC; Akaike, 1974) and the *Bayesian information criterion* (BIC; Schwarz, 1978), sometimes also known as the Schwarz information criterion. The AIC is defined as

$$\text{AIC} = -2L(y|x,\theta) + 2k, \tag{1.8}$$

where k is the number of parameters estimated in a model, and the BIC is defined as

$$\text{BIC} = -2L(y|x,\theta) + k\log(N), \tag{1.9}$$

where N is the sample size. Both the AIC and BIC penalize a model by its number of parameters (i.e., its complexity). The lower the information criterion score, the better the model fit is relative to the model's complexity. The BIC penalizes model complexity more severely than the AIC by including sample size as a factor, and some researchers prefer the BIC to the AIC because they believe the BIC is more likely to help them avoid selecting an overfitted model. On the other hand, while the AIC actually is based on information theory, the BIC is not. Vrieze (2012) presents a thoughtful discussion of alternative information criteria.

In addition to computing overall measures of fit with the data, researchers are well advised to evaluate the model in greater detail. Broadly speaking, there are three ways that this can be done: investigating the extent to which a candidate model fits all of the observations equally well, ascertaining whether particular observations have a disproportionate influence on the model parameter estimates, and assessing how sensitive the model fit is to perturbations of the parameter estimate values. These are related but distinct investigations. Assessments of how well a model fits individual observations typically are conducted using residuals, which are based on the difference between an observation's value and the model's prediction thereof. Assessments of the influence an observation has on the model parameter estimates usually are done via "leverage" statistics, which measure the sensitivity of parameter estimates to the presence or perturbation of the observation. Assessments of model fit sensitivity to perturbations of model parameter values are less common than the other two types of assessment but can be done as a by-product of the other two kinds or simply by computing likelihoods of the model with alternative parameter values sampled within their respective standard error or confidence bounds.

Turning first to residuals, four commonly employed kinds are the raw (or response), Pearson, Anscombe, and deviance residuals. The response residuals are just the difference between the observations and the model's predictions:

$$r_i = y_i - y_i' \big| \left(x_i, \hat{\theta} \right). \tag{1.10}$$

The *Pearson residuals* scale the raw residuals by the standard error of the predicted value:

$$r_{pi} = \frac{r_i}{\sqrt{V \left(y_i' \left(x_i, \hat{\theta} \right) \right)}}, \tag{1.11}$$

where V is the variance function associated with the GLM's distribution. Pearson residuals are asymptotically normally distributed under appropriate conditions, but in real situations, they can be strongly skewed. *Anscombe residuals* (Anscombe, 1953) transform the numerator and denominator of the Pearson residuals to make them unimodal and symmetric, thereby as close to a normal distribution as possible. A full explanation of these residuals would constitute a digression here, but Gill (2000) provides an accessible and detailed explanation.

A different type of residual is the deviance residual. The *deviance* statistic associated with a model is defined as twice the difference between the log-likelihood of a "saturated" model that perfectly predicts the observations y_i and the log-likelihood of the model:

$$D(y|x, \theta) = 2 (L(y|\xi) - L(y|x, \theta)), \tag{1.12}$$

where ξ is the vector of parameters that perfectly predicts the y_i. The *deviance residual*, then, is

$$r_{di} = \text{sign}(r_i) \sqrt{d_i(x_i, \theta)}, \tag{1.13}$$

where sign() takes the sign of the term in the parentheses, and d_i denotes the ith contribution to the deviance,

$$d_i(y_i|x_i, \theta) = 2 (L(y_i|\xi_i) - L(y_i|x_i, \theta)). \tag{1.14}$$

In addition to examining residuals, assessing the effect of individual observations on the model provides a second way of evaluating the model. A commonplace assumption among researchers is that excessively influential observations also are outliers, but this is not always true. Thus, there is a role for *influence statistics*, which measure the impact that an observation has on the model. The main idea underpinning influence statistics is cross-validation, whereby one observation at a time is removed from the data and the model is fitted to the remaining data. The most popular influence statistic is Cook's distance, which measures the sum of the changes in regression coefficients when one observation is removed from the data (Cook, 1977). Cook's distance was derived in the context of linear regression, and variants of it have subsequently been developed for other GLMs. Likewise, a popular standardized measure of change in a model parameter when an observation is removed is the change in the coefficient divided by the standard error of the original parameter estimate, for example,

$$\Delta_i \left(\hat{\beta} \right) = \left(\hat{\beta} - \hat{\beta}_{(i)} \right) \Big/ \hat{\sigma}_{\hat{\beta}}, \tag{1.15}$$

where $\hat{\beta}_{(i)}$ denotes the ML estimate of β when the ith observation has been removed. These influence statistics are called "dfbetas."

1.4   Examples

Having reviewed GLMs and introduced some ideas about bounds, we will finish our introduction with two examples. We hope these examples will make these concepts more concrete and also illustrate some of the benefits from using techniques that take bounds into account.

1.4.1   *Absolute Bounds Example*

Income distributions are a prototypical example of a distribution with one bound. We will use household income data for households with positive income for two of the years, 2010 and 2015, from the American Community Survey database (U.S. Census Bureau, 2015), for our example. These data comprise 148,076 households, and their income distribution and normal quantile-quantile (Q-Q) plot are shown in Figure 1.1. The distribution is strongly skewed (the largest household income exceeds \$2 million), and the Q-Q plot shows that it deviates far from a normal distribution.

Figure 1.1   Household Income Distribution and Q-Q Plot

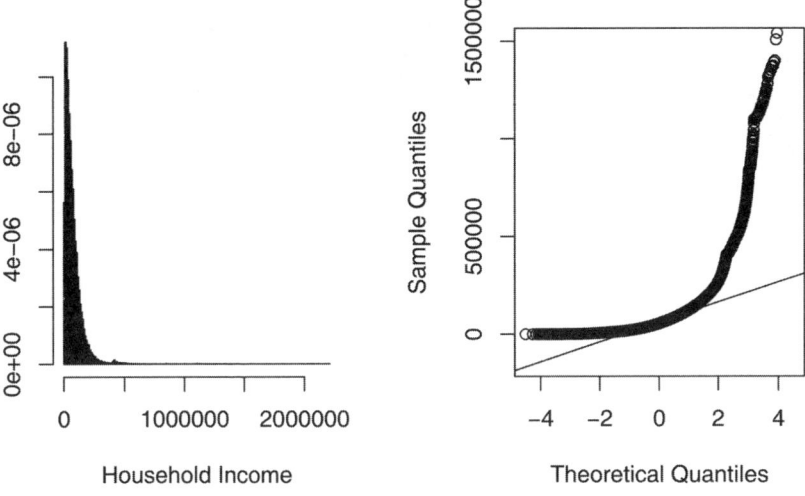

Figure 1.2   Log(Household Income) Distribution and Q-Q Plot

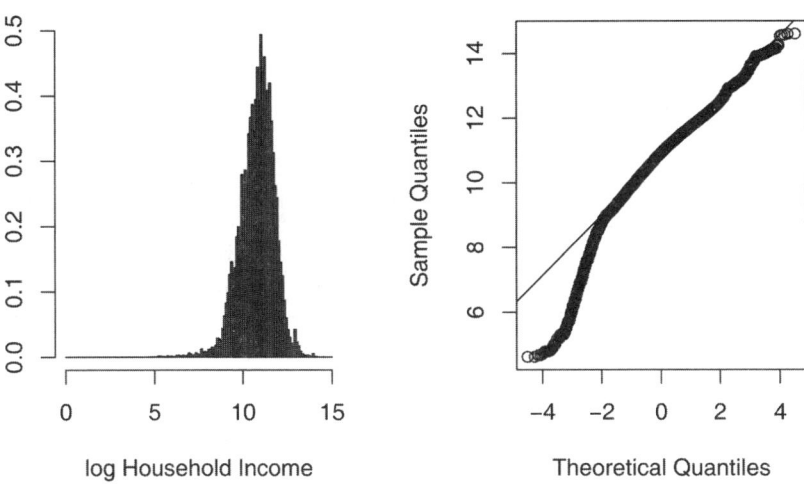

The distribution of the log income and its Q-Q plot in Figure 1.2 suggest that the lognormal may be an appropriate distribution for modeling these data. The Q-Q plot shows that the log income distribution closely corresponds to the normal distribution for all but approximately 2% of

the data in the lower tail (i.e., the data from about 2 standard deviations below the mean downward).

Suppose we wish to compare the incomes for households that obtained food stamps with those that did not, and we also would like to ascertain whether the difference between them changed from 2010 to 2015. A linear regression model (equivalent to an analysis of variance [ANOVA] with Type III sums of squares) yields this outcome:

$$Y|x, \beta \sim N\left(\mu, \sigma^2\right)$$
$$\mu = x\beta = 75990.63 - 54389.33x_1 + 10029.23x_2 - 4476.95x_1x_2,$$
$$(1.16)$$

where x_1 is the food-stamp dummy variable ($x_1 = 1$ for households obtaining food stamps and 0 for households with no food stamps), x_2 is the year dummy variable ($x_2 = 1$ for 2015 and 0 for 2010), and x_1x_2 is the product of x_1 and x_2.

Now, there are two commonplace versions of a "lognormal" model among researchers. One of these is a linear regression model with the log of the dependent variable. Let us denote this as the "log-DV" model. This model with the same predictors yields this outcome:

$$\log(Y)|x, \beta \sim N\left(\mu, \sigma^2\right)$$
$$\mu = x\beta = 10.888 - 1.223x_1 + 0.101x_2 + 0.053x_1x_2 \qquad (1.17)$$
$$\log(\sigma) = \delta = -0.101.$$

All of the coefficients are statistically significant, not least because of the enormous sample size. Our focus here is on the structure, effects, and goodness of fit of each model. Beginning with the structure, we can see from the coefficients that there is one obvious difference between the two models, namely that the linear model has a negative interaction term and the log-DV model has a positive interaction term. It also should be borne in mind that the magnitudes of the coefficients differ considerably due to the fact that one model is in the linear scale and the other is in the log scale. There are several more or less equivalent ways to compare the effects of these two models, all of which require transforming from one model's scale to the other's. We will transform the log-DV model's estimates to the linear scale.

Table 1.1 displays the predicted means for each of the models in their respective scales. In the row below the log-DV model's means, the corresponding means have been computed in the linear scale. Recall that although $E(\log(Y)) = \mu$, the expectation of Y is $E(Y) = \exp(\mu + \sigma^2/2)$.

Table 1.1   Linear and Log-DV Model-Predicted Means and Effects

	Food Stamp Year	No 2010	No 2015	Yes 2010	Yes 2015
Linear model $\hat{\sigma} = 78,270$	Means	75,990.63	86,019.85	21,601.30	27,153.58
			Ratios	3.52	3.17
Log-DV model $\hat{\sigma} = 0.9036$	Means	10.8878	10.9883	9.6652	9.8188
	Linear scale	80,505.23	89,012.98	23,705.75	27,642.44
			Ratios	3.40	3.22

For example, the 2010 no-food-stamps mean is $E(Y) = \exp(10.8878 + 0.9036^2/2) = 80,505.23$ (allowing for a small roundoff error). The estimated standard deviation, $\hat{\sigma}$, for each model is displayed in the leftmost column. Although the log-DV model's transformed means differ from those of the linear model, the effect sizes are fairly similar. The linear model's mean of the no-food-stamp households in 2010 is 3.52 times higher than the food-stamp household mean for the same year $(75,990.63/21,601.30 = 3.52)$, and the corresponding log-DV ratio of its means is 3.40. For 2015, the linear model means ratio is 3.17 and the log-DV means ratio is 3.22.

Now we turn to goodness of fit. Both models perfectly reproduce the sample means in their respective scales. That is, the linear regression model perfectly reproduces the means in the original scale, and the log-DV GLM perfectly reproduces the means in the log scale. However, as we saw earlier, the log-DV model does not produce the same mean in the linear scale as the linear model does (e.g., the mean income for households not receiving food stamps in 2010 is 80,505.23 rather than 75,990.6).

Now let us turn to the second version of a lognormal model, in which the link function for the mean response is the log. This model yields a somewhat different set of coefficients from those in the log-DV model of equation (1.17):

$$Y|\boldsymbol{x}, \boldsymbol{\beta} \sim LN\left(\mu, \sigma^2\right)$$
$$\log(\mu) = \boldsymbol{x}\boldsymbol{\beta} = 11.2384 - 1.2579x_1 + 0.1240x_2 + 0.1048x_1x_2.$$
$$(1.18)$$

We have deliberately chosen a "trivial" example to illustrate the fact that transforming the lognormal model's estimates back to the linear scale reproduces the sample means in the linear scale, just as the linear model does. For example, using a more precise estimate of the intercept to diminish roundoff errors, the model's estimated mean income for households not receiving food stamps in 2010 is $\exp(11.238365) = 75,990.6$. Moreover, the lognormal model has exactly the same log-likelihood as the linear model. Thus, the most important practical difference between the log-DV and lognormal models is that the log-DV model rescales the dependent variable, whereas the lognormal model rescales the model.

1.4.2 Censoring Bounds Example

We now turn to an example of a model for a censored dependent variable. A class of 159 second-year Australian National University psychology students completed the DOSPERT, and we will model their responses on the "ethical risk-taking" subscale. Items in this subscale ask respondents to rate their likelihood to commit acts such as cheating on an exam and pirating software. A substantial number of the students' responses yielded the lowest possible score. Why is this lowest possible score "censored"? One way to think about this is that it may be possible to find other examples of unethical acts that even some of these 20 respondents would rate themselves as likely to do. After all, the DOSPERT includes only eight examples of unethical acts, nowhere near exhausting the human repertoire.

Figure 1.3 displays a histogram of the subscale responses, showing 20 cases on the lower boundary of the subscale. Two fitted distributions also are displayed. The dashed-line distribution is a normal distribution that reproduces the sample mean, $\hat{\mu} = 15.12$, and standard deviation, $\hat{\sigma} = 5.48$. The solid-line distribution is a normal distribution fitted via a *Tobit model*, which takes censoring into account (Tobin, 1958). It estimates the mean as $\hat{\mu} = 14.73$ and standard deviation as $\hat{\sigma} = 6.11$, appropriately decreasing the mean and increasing the standard deviation estimates. Moreover, the Tobit model's distribution fits the data better than the normal distribution: The log-likelihood for the Tobit is -471.1 whereas for the normal, it is -495.6.

The Tobit model is the most popular GLM for censored outcomes. It assumes that the data are sampled from a censored normal distribution. In our example, that means assuming that the 20 boundary cases, if they were uncensored, would be distributed as in the lower tail of a normal distribution whose support extends beyond the scale's boundary (the lower tail of the solid-line distribution in Figure 1.3). Censoring can occur at either end of a scale (or both), and there are several kinds of censoring. These matters will be elaborated in Chapter 6, but here

Figure 1.3 Normal and Tobit Distributions Fitted to Subscale Scores

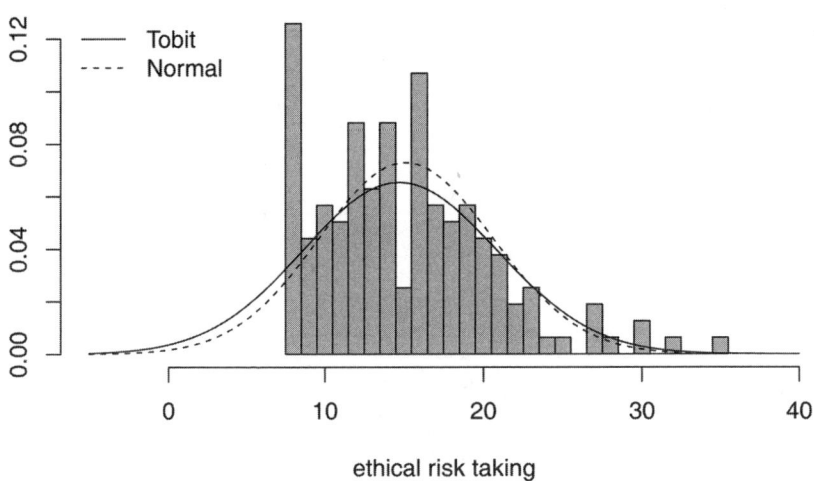

we focus on a lower-censored Tobit model. The traditional notation for such a model describes the uncensored observations in terms of equation (1.1), that is,

$$y_i = x_i\beta + \epsilon_i, \tag{1.19}$$

where $\epsilon_i \sim N(0, \sigma)$. For the variable's censored observations, suppose τ is the censoring threshold. Then

$$x_i\beta + \epsilon_i \leq \tau, \tag{1.20}$$

so that $\epsilon_i \leq \tau - x_i\beta$ and therefore

$$\Pr(y_i \leq \tau \,|x_i) = 1 - \Phi((x_i\beta - \tau)/\sigma_i), \tag{1.21}$$

where Φ is the standard normal cumulative distribution function. A model for the censoring rate is given by equation (1.21).

We have seen in our example that the Tobit model gives different estimates of the mean and standard deviation from those of the linear regression model. The linear regression model's estimate of the mean is just $\mu_x\beta$, where μ_x denotes the vector mean of the predictor variables. The Tobit model's mean estimate is

$$E[y] = \Phi_\tau\mu_x\beta + \sigma\phi_\tau + \tau(1 - \Phi_\tau), \tag{1.22}$$

where $\Phi_\tau = \Phi((\mu_x\beta - \tau)/\sigma)$ and $\phi_\tau = \phi((\mu_x\beta - \tau)/\sigma)$. We now consider a linear regression and a Tobit model with scores on the DOSPERT

Table 1.2   Linear Regression and Tobit Model Summaries

Model	Coefficient	Estimate	Standard Error		
Regression	Intercept	1.414	1.274	Log-likelihood	−449.4
	Health risk	0.687	0.062	t	11.132
Tobit	Intercept	−0.592	1.464	Log-likelihood	−424.4
	Health risk	0.770	0.070	t	10.968
Quantiles	10%	25%	50%	75%	90%
Empirical	8	11	14	18	22
Regression	10.35	12.41	14.48	17.22	19.42
Tobit	9.41	11.72	14.03	17.11	19.57

health subscale predicting scores on the ethical risk-taking subscale. The two models' coefficients, t statistics, predicted versus empirical quantiles, and log-likelihoods are displayed in Table 1.2. The linear regression model's coefficient is smaller than the Tobit's, although their t values are similar (due to the Tobit model's larger residual standard error estimate). As before, the log-likelihoods indicate that the Tobit model is the better fit. This also is borne out by the predicted quantiles, which in all but one case are closer to their empirical counterparts for the Tobit than for the regression model.

In this chapter, we have tried to set the stage for the rest of the book by reviewing the general linear model, introducing the concepts needed for understanding the nature of bounds on variables, and presenting what we hope are motivating demonstrations of models for bounded variables. We now move on to considering models for singly bounded variables, when the boundary is treated as absolute.

CHAPTER 2. MODELS FOR SINGLY BOUNDED VARIABLES

In this chapter, we focus on models for variables with nonnegative values, as these are typical of singly bounded variables. However, nearly all of the material in this chapter generalizes to variables with lower or upper bounds that are not zero. We start with the lognormal distribution because of its familiarity to readers, but also because it illustrates how a bound induces dependency between location and dispersion. The chapter then introduces models using the popular gamma and Weibull distributions and provides two examples of their application to real data. The second section describes popular model diagnostic tools and illustrates their use. The final section deals with the often neglected issue of how to treat cases that are on the boundary, with an example illustrating the kinds of decisions the researcher may have to make regarding models for boundary cases.

2.1 GLMs for Singly Bounded Variables

2.1.1 *Alternative Distribution Models*

The collection of distributions for modeling singly bounded random variables is abundant and rich (Marshall & Olkin, 2007), with a lengthy history due to the preponderance of singly bounded variables in well-researched domains throughout the sciences and engineering. We are limiting our discussion to just three of the most popular: the lognormal, gamma, and Weibull distributions. We have selected them because they are important, they have been made widely available in statistical packages, and their properties are relatively easy to understand.

Although the three distributions introduced in this section have distinct properties, they often fit data similarly well, thereby sometimes making it difficult for researchers to choose one over the others on the basis of model fit alone. We shall see this in both examples presented in the subsections following this one. Moreover, these distributions also are able to closely mimic the shape of a normal distribution. Thus, researchers can sometimes "get away with" linear normal-theory regression models even with variables that have a bound, usually when the bulk of the data is far from the bound. One advantage of this quandary is that if these alternative distribution models yield similar results in terms of covariate effects and model diagnostics, then it suggests that the findings are robust. However, all of this underscores the problematic fact that in many applications of these distributions throughout the human sciences, little attention is given to their appropriateness for modeling the actual psychological, social, economic, or political processes generating the data. At the very least, then, researchers are advised to try fitting each of these distributions to their data, so that they have some basis for claims regarding the solidity of their models and findings.

We now turn to brief descriptions of the lognormal, gamma, and Weibull distributions and their respective conceptual bases. Their descriptions here will be limited to versions with support on the nonnegative half of the real line, and researchers should bear in mind that these versions can be adapted to other support ranges via the usual linear scale transformation (e.g., a range whose lower bound is greater than zero, such as reaction times).

The probability density function (pdf) of the two-parameter *lognormal distribution* conventionally is written as

$$f(x, \mu, \sigma) = \frac{\exp\left(-(\log x - \mu)^2 \big/ 2\sigma^2\right)}{\sigma x \sqrt{2\pi}}, \tag{2.1}$$

where $x \geq 0$, $\mu > 0$, and $\sigma > 0$. The fact that these two parameters are borrowed from the normal distribution somewhat obscures their true nature. As we pointed out in Chapter 1, although the expected value of $\log(X)$ is μ and its variance is σ^2, the expected value of X itself is $\exp(\mu + \sigma^2/2)$ and its variance is $(\exp(\sigma^2) - 1)\exp(2\mu + \sigma^2)$. Thus, the mean and variance of X are not independent of one another. A model of μ therefore accounts for conditional variability in the mean and variance that is not accounted for by a model of σ and vice versa. More details regarding the characteristics of the lognormal distribution are available in Chapter 14 of Johnson, Kotz, and Balakrishnan (1994). In describing such models, researchers will need to be careful to specify their meanings when they refer to location and dispersion submodels. We described the GLM for the lognormal in Chapter 1. The link functions for μ and σ are the log.

As the central limit theorem tells us, the normal distribution is a limiting distribution for sums, and therefore the lognormal is a limiting distribution for products. Thus, a natural justification for using the lognormal distribution is when we want to estimate a multiplicative model. The classic application of this kind is in economics, epidemiology, and biology, motivated by considerations of what has been termed *proportional effects*. For example, the effect of a medication on a disease may be proportional to the rate of progression of the disease. Via the lognormal distribution, a multiplicative model can be transformed into an additive model and vice versa.

The pdf of the two-parameter *gamma distribution* is

$$f(x, \lambda, \nu) = \lambda^{-\nu} x^{\nu-1} \exp\left(-x/\lambda\right) \big/ \Gamma(\nu), \tag{2.2}$$

where $x \geq 0$, $\lambda > 0$, $\nu > 0$, and $\Gamma(\cdot)$ is the gamma function. In this version of the distribution, ν is a shape parameter and λ is known as a rate parameter. The mean is $\mu = \nu\lambda$ and the variance is $\sigma^2 = \nu\lambda^2$, so clearly the mean and variance are not independent of each other. Available literature and software include alternative parameterizations of the gamma distribution, the most common of which uses the reciprocal of λ (Marshall & Olkin, 2007). The GLM for the gamma distribution conventionally reparameterizes the distribution and uses the log as the link function for μ and σ.

When $\nu = 1$, the gamma distribution becomes the exponential distribution, and when $\nu = \kappa/2$ and $\lambda = 2$, it becomes the chi-square distribution with κ degrees of freedom. The chi-square special case links the gamma distribution with the sum of squared standard normal random variables. A more important characteristic of gamma random variables is the *reproductive property*: The sum of J independent gamma random variables, X_j, that have the same λ parameter but possibly differing ν_j is itself a gamma random variable with parameters λ and $\sum \nu_j$ (N. Johnson et al., 1994). The reproductive property renders the gamma distribution suitable for modeling variables that may be considered sums of this kind. Examples of such an application may be found in eye-tracking studies involving models of sums of fixation durations, as foreshadowed by Aribarg, Pieters, & Wedel (2010) and elaborated in Smithson & Shou (2014).

The pdf of the two-parameter *Weibull distribution* is

$$f(x, \eta, \kappa) = \eta x^{\eta-1} \exp\left(-(x/\kappa)^\eta\right) \big/ \kappa^\eta, \tag{2.3}$$

where $x \geq 0$, $\eta > 0$, and $\kappa > 0$. The mean is $\mu = \kappa \Gamma\left(1 + 1/\eta\right)$ and the variance is $\sigma^2 = \kappa^2 \left(\Gamma\left(1 + 2/\eta\right) - \Gamma^2\left(1 + 1/\eta\right)\right)$. The literature and software include several alternative parameterizations of the Weibull distribution (Hallinan, 1993), so researchers must take care to ascertain which is being employed in a given article or computing environment. Like the gamma distribution, the Weibull includes the exponential distribution as a special case and can be interpreted as an exponential distribution with power and scale parameters added to it. The Weibull is popular in areas such as reliability engineering or life expectancy estimation, because of its versatility and the fact that it is related to several other probability distributions.

2.1.2 Time Required to Start a Business by Nation

Our first example uses World Bank data on various performance indicators of 190 national economies. One of these is the average number of days required to start a business. This is an indicator of how feasible doing business is in each nation and therefore an indicator of the potential for national socioeconomic development. A recent World Bank report (Coste, Meunier, Novik, Reeves, & Tjong, 2018) linked ease of starting a business with indicators related to business registry information transparency and control of corruption. The Worldwide Governance indicator, control of corruption, is based on perceptions of the degree to which public power is exercised for private gain, including control of the state by elites and private interests. Higher scores on the control of corruption index indicate less corruption. In the report, the effect of business transparency on time required to start a business was identified by controlling for gross domestic product (GDP) per capita. However, this analysis did not take into account more proximal variables such as the amount of "red tape," or bureaucratic procedures, required to start a business.

We therefore reassess the effect of corruption control on business startup time, taking red tape into account via another indicator in the World Bank database, namely, number of procedures required to start a business. Data for all three indicators are available for 172 nations (the business startup time and required number of procedures data are for 2017, while the corruption index data are for 2016). Number of procedures required and the corruption index are modestly negatively correlated ($r = -0.358$), suggesting that greater corruption goes with greater numbers of required procedures. Figure 2.1 displays the distribution of the average required startup days for each country. The distribution is strongly skewed with an obvious outlier (Venezuela, at 280 days).

We fit lognormal, gamma, and Weibull models to the data and compare these with a Gaussian linear regression model. There seems to be no compelling theoretical rationale for preferring any of these distributions over the others. It could be argued, for example, that the

Figure 2.1   Average Days Required to Start a Business

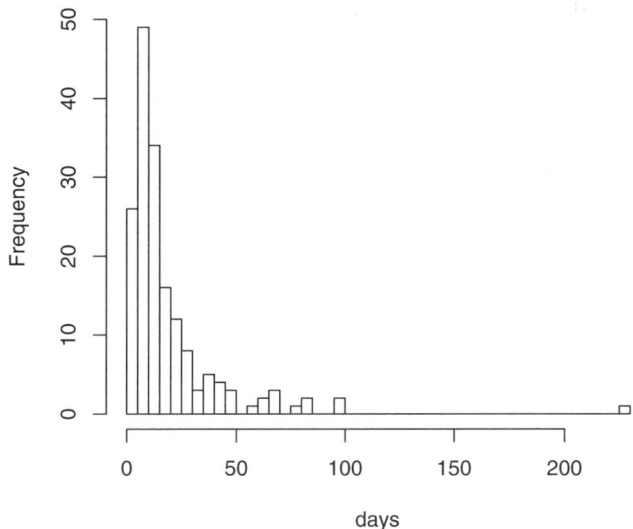

collection of bureaucratic and financial procedures needed to start a business operates as an ordered sequence, each requiring its own time for completion, so that the total time required is a sum of these component times. This assumption would favor the gamma model because of its reproductive property, but there is no evidence for the assumption, and counterarguments are readily available.

Neither GDP per capita nor gross national income (GNI) per capita contribute to model fit, so the final models have just two predictors, the "proximal" predictor (number of procedures required to start a business) and "distal" predictor (corruption index). Likewise, an interaction term for the two predictors does not improve model fit, nor does including the predictors in the dispersion submodels. Table 2.1 displays the model coefficients for these models, using standardized-score versions of the predictors, so that the coefficients can be compared across different models.

In all models, the predictors contribute independent effects whose coefficients have opposite signs. As would be expected, a greater number of required procedures increases the required startup time, while greater control over corruption decreases startup time. The findings in the lognormal, gamma, and Weibull models support the hypothesis that the corruption index predicts business startup time even when the

Table 2.1 Gamma, Weibull, Lognormal, and Gaussian Models for Business Startup Time

	Coefficient	Estimate	SE	t	p
Gamma model					
Intercept	β_0	2.694	0.044		
Number of procedures	β_1	0.649	0.052	12.420	<.0001
Control of corruption	β_2	−0.179	0.046	−3.881	.0002
	$\log(\sigma)$	−0.546	0.051		
Weibull model					
Intercept	β_0	2.704	0.048		
Number of procedures	β_1	0.657	0.060	11.004	<.0001
Control of corruption	β_2	−0.170	0.048	−3.537	.0005
	$\log(\sigma)$	0.494	0.054		
Lognormal model					
Intercept	β_0	2.681	0.080		
Number of procedures	β_1	0.525	0.030	17.555	<.0001
Control of corruption	β_2	−0.303	0.081	−3.751	.0020
	$\log(\sigma)$	−0.553	0.054		
Linear regression model					
Intercept	β_0	19.385	1.424		
Number of procedures	β_1	15.053	1.541	9.770	<.0001
Control of corruption	β_2	−2.493	1.541	−1.618	.1070

number of required procedures for setting up a business is taken into account.

The linear regression model does not fit the data as well as the other three models. The correlation between the dependent variable and the linear model predictions is 0.653, whereas for the gamma, lognormal, and Weibull models, the correlations are 0.781, 0.780, and 0.787, respectively. The linear regression model also fails to identify a significant corruption index effect. Clearly, a linear Gaussian model is not appropriate for this data set.

While the correlations between the dependent variable and model predictions are nearly identical for the three models, nevertheless, there are some noticeable differences in the predicted values between the lognormal and the other two models. Table 2.2 shows the empirical quantiles of average business startup times for the nations and compares

Table 2.2   Empirical Quantiles Versus Quantile Gamma, Weibull, and Lognormal Fitted Values

Quantiles	0.1	0.25	0.5	0.75	0.9
Empirical	4.50	6.50	12.00	22.13	40.95
Gamma	6.53	8.92	14.54	21.45	40.02
Weibull	6.61	8.98	14.62	21.59	40.65
Lognormal	6.16	10.12	13.91	21.92	36.21

these with the quantiles of the three models' predicted values. The gamma and Weibull models have more accurate quantiles than the lognormal model, but all three models have larger quantiles than the data in the lower half of the distribution.

2.1.3   Performance in a Stroop Task Experiment

Harris, Harris, and Miles (2017) report an investigation of the effect of self-affirmation on two aspects of performance that have been related to executive functioning: working memory and inhibition. Inhibition was assessed by a Stroop task. The Stroop task required participants to indicate the color of a string of letters presented on a computer screen. There were three types of letter-strings: "neutral" ("XXXX" colored red or blue), "congruent" ("Red" colored red or "Blue" colored blue), and "incongruent" ("Red" colored blue or "Blue" colored red). Participants completed 60 trials (20 of each type, in random order). The amount of time it took participants to identify the color (response time) was one of the dependent variables.

Eighty-three participants were randomly assigned to either a self-affirmation task or a control task and then completed computerized tasks that included the Stroop. In addition, measures were taken on several personality and individual-difference traits, including optimism as measured by the Life Orientation Test–Revised (LOT-R). We will analyze the joint impact of the self-affirmation manipulation and trait optimism on response times in the Stroop task. The guiding hypothesis is that self-affirmation should decrease response times by making participants more self-confident.

In the absence of a compelling argument that favors any one distribution, we fit lognormal, gamma, and Weibull models to the data and compare these with a Gaussian linear regression model. Again, the correlations between the dependent variable and model predictions are very

Table 2.3　Gamma, Weibull, and Lognormal Models for Stroop Task
Response Times

Gamma model					
	Coefficient	Estimate	SE	t	p
Intercept	β_0	5.372	0.332		
Affirmation	β_1	−1.791	0.517	−3.467	.0009
LOT-R	β_2	0.065	0.100	0.647	.5198
Affirmation * LOT-R	β_3	0.492	0.163	3.022	.0034
	$\log(\sigma)$	−0.684	0.075		
Weibull model					
	Coefficient	Estimate	SE	t	p
Intercept	β_0	5.397	0.260		
Affirmation	β_1	−1.908	0.434	−4.401	<.0001
LOT-R	β_2	0.060	0.077	0.775	.4408
Affirmation * LOT-R	β_3	0.526	0.137	3.843	.0002
	$\log(\sigma)$	0.815	0.088		
Lognormal model					
	Coefficient	Estimate	SE	t	p
Intercept	β_0	5.149	0.376		
Affirmation	β_1	−1.903	0.594	−3.203	.0020
LOT-R	β_2	0.133	0.111	1.197	.2348
Affirmation * LOT-R	β_3	0.525	0.168	3.121	.0025
	$\log(\sigma)$	−0.606	0.078		

Note: LOT-R = Life Orientation Test–Revised.

similar (0.507, 0.501, and 0.501, respectively). A linear regression model
again performs worse (correlation is 0.472).

Table 2.3 displays the coefficients for the final three models (we
subtracted 250 ms from the response times to enable the data to be
treated as though the lower boundary is 0). As in the previous example,
the effects are quite similar across all three models. There were no
discernible effects of either self-affirmation or LOT-R in the dispersion
submodel, but the location submodel reveals a consistent interaction

effect. One interpretation of it is that optimism has an effect only in the self-affirmation condition. In that condition, greater optimism increases response time. Moreover, the increase is sufficient to eliminate the impact of self-affirmation for participants scoring far enough above the mean of LOT-R (the mean is approximately 3.12 and the affirmation effect goes to 0 when LOT-R $= -\beta_1/\beta_3$).

We may compare the three models' predictions of the mean response times in the two experimental conditions. The mean LOT-R scores for the two conditions differ slightly due to randomized assignment (3.20 for the no-affirmation and 3.02 for the affirmation conditions). Taking these into account, for the no-affirmation versus affirmation conditions, the gamma model's predicted response times are 515.13 and 443.11 ms, the Weibull model's predictions are 517.70 and 442.23 ms, and the lognormal model's predictions are 514.07 and 437.49 ms. The empirical means are 515.47 and 455.52 ms. The gamma model is arguably the most accurate here, although all three models underestimate the affirmation condition mean.

2.2   Model Diagnostics

The usual model diagnostics for GLMs may be applied in evaluating models for singly bounded random variables. These include plots of residuals and influence statistics, cross-validation, and the parameter estimation correlation matrix. The latter can be especially relevant when there are many observations close to the boundary, as that is when high correlations among parameter estimates are most likely to occur. To illustrate model diagnostics, we focus on the World Bank business startup time example.

A Cook's distance plot in Figure 2.2 from the gamma model reveals that the outlier in startup time (the observation furthest to the right along the log(fitted) axis, at log(280) $= 5.63$) is not a strongly influential observation. Instead, the influential observations are in the middle of the distribution. This is a demonstration of the fact that highly influential observations need not be outliers of the dependent variable. A similar pattern is observable for the Weibull and lognormal models.

Finally, Table 2.4 shows that the parameter estimate correlation matrix for these models does not yield any worrisome values, with the strongest correlation being about 0.4, between the coefficients for the two predictors.

28

Figure 2.2   Gamma Model Cook's Distance Plot

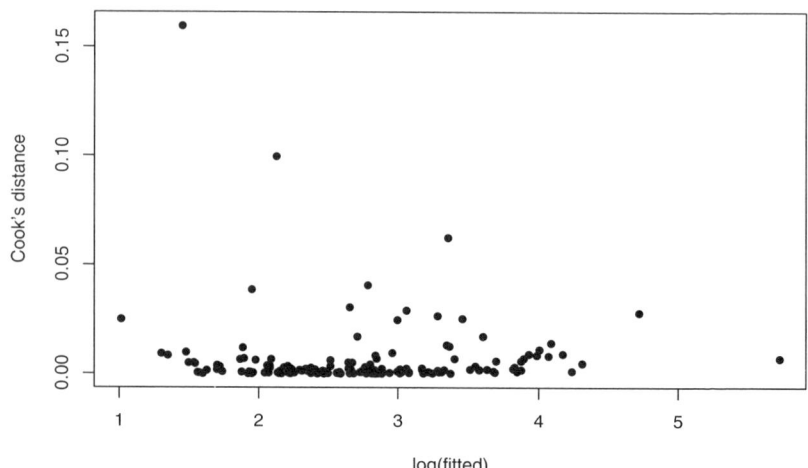

log(fitted)

Table 2.4   Parameter Estimation Correlation Matrices

Gamma	β_0			
	−.00007	β_1		
	−.00001	.36490	β_2	
	−.00001	−.00016	−.00003	σ
Weibull	β_0			
	.00259	β_1		
	.00704	.41476	β_2	
	.25681	.01011	.02743	σ
Lognormal	β_0			
	−.36475	β_1		
	.44195	.54306	β_2	

2.3 Treatment of Boundary Cases

Boundary cases can pose problems for conventional models of singly bounded random variables, and the choice of treatment for them is an important decision. Boundary cases often occur in financial or economic applications. For instance, many insurance policies in an insurance company's client list or lines of business in a bank may show zero loss in any given time period. Alternatively, distributions may exhibit an inflated spike at or near the boundary even though the bulk of the distribution is far from the boundary. Most human population life span distributions have this characteristic, with a noticeable peak near zero due to infants who die during their first year, but a mode that is far into the adult age range.

Singly bounded variables present the analyst with two alternative ways of construing boundary cases common to all bounded variables: treating the boundary as a matter of censoring or truncation, or treating it as a true score and therefore an absolute bound. If the boundary is regarded as absolute, then the boundary cases have to be treated as true scores. For variables bounded below by zero, this creates a problem for GLMs using distributions such as the lognormal, gamma, or Weibull. Most readers will recall that log(0) is undefined, so such models cannot handle zeros. If there are not many zeros or if the zeros result from a roundoff (e.g., all observed values below some threshold are rounded down to 0), then a practical (and popular) approach is to add a small constant to the zeros, such as 0.001. If a researcher chooses this alternative, then it is a good idea to investigate the sensitivity of the model to the size of this constant (e.g., do the results change if the constant is 0.005 instead of 0.001?).

If there are a lot of zeros and these are genuinely true scores, then there are two kinds of models available that again are common to all bounded variables and a third model that (to our knowledge) is available only for variables with a single lower bound. The two common kinds of models are so-called hurdle and boundary-inflated models. In some of the research literature, this terminology has been inconsistently used (more on this in later chapters), but in this book, we aspire to consistent usage wherever possible. A *hurdle model* is essentially a two-stage model, with an initial model distinguishing between cases on and off the boundary and a second model for the cases that are not on the boundary (Mullahy, 1986). The first model usually is a logistic regression, but in general, it posits a Bernoulli (boundary) random variable to account for the cases falling at the boundary. We may write the hurdle model for a lower-bounded variable as

$$P(Y = y) = \begin{cases} \pi & \text{if } y = \tau \\ (1 - \pi)f(y) & \text{if } y > \tau, \end{cases} \tag{2.4}$$

where τ is the boundary value of y and $0 \leq \pi \leq 1$. In the components of this model, $f(y) = h^{-1}(x\boldsymbol{\beta})$, and $\pi = \text{logit}^{-1}(z\boldsymbol{\gamma})$, where h is a link function, x and z are predictors, and $\boldsymbol{\beta}$ and $\boldsymbol{\gamma}$ are coefficients.

A *boundary-inflated model* is a mixture model, with one mixture component, $f(y)$, covering the entire range of the variable (including the boundary) and the other component modeling the "excess" cases at the boundary (Lambert, 1992). So, a key difference between this model and the hurdle model is that a boundary-inflated model has an "uninflated" predicted density at the boundary, whereas the hurdle model does not. The boundary-inflated model for a lower-bounded variable may be written as

$$P(Y = y) = \begin{cases} \pi + (1 - \pi)f(\tau) & \text{if } y = \tau \\ (1 - \pi)f(y) & \text{if } y > \tau. \end{cases} \tag{2.5}$$

Note that in order for a boundary-inflated model to be available, $f(\tau)$ must be defined. For example, if $\tau = 0$, then it is possible to construct boundary-inflated models for gamma and Weibull distributions under special conditions but not for the lognormal distribution. Even so, zero-inflated models for gamma and Weibull distributions are not of much practical value because even when their densities at zero are defined, they generally are 0.

For the third type of model, we regard the random variable as *semicontinuous* (i.e., a variable whose distribution is continuous with a point mass at the boundary). Note that this is not a mixture model in the sense we have used just before, because this is a single distribution rather than a mixture of two distributions. An appropriate semicontinuous distribution that has been applied in domains such as actuarial science, economics, medicine, and ecology is a special case of the Tweedie distribution family (Tweedie, 1984). Tweedie distributions include several members of the exponential distribution family. Their density function does not have an analytical expression, but for our purposes, they may be described as a random variable Y having a mean, μ, and variance function V such that $V(\mu) = \mu^{\omega}$ and $V(Y) = \phi\mu^{\omega}$, where ω is any non-negative value that is not in the $(0, 1)$ interval and $\phi > 0$ is a dispersion parameter. When $1 < \omega < 2$, Y follows a compound Poisson-gamma distribution and therefore has a point mass at 0. In the next subsection, we provide an example applying the Tweedie distribution.

2.3.1 Ambulance Arrival Times

Welsh regional health boards present data from their respective regions on the time it takes for an emergency-call ("Red") ambulance to arrive

Figure 2.3 Welsh Regional Ambulance Travel Times

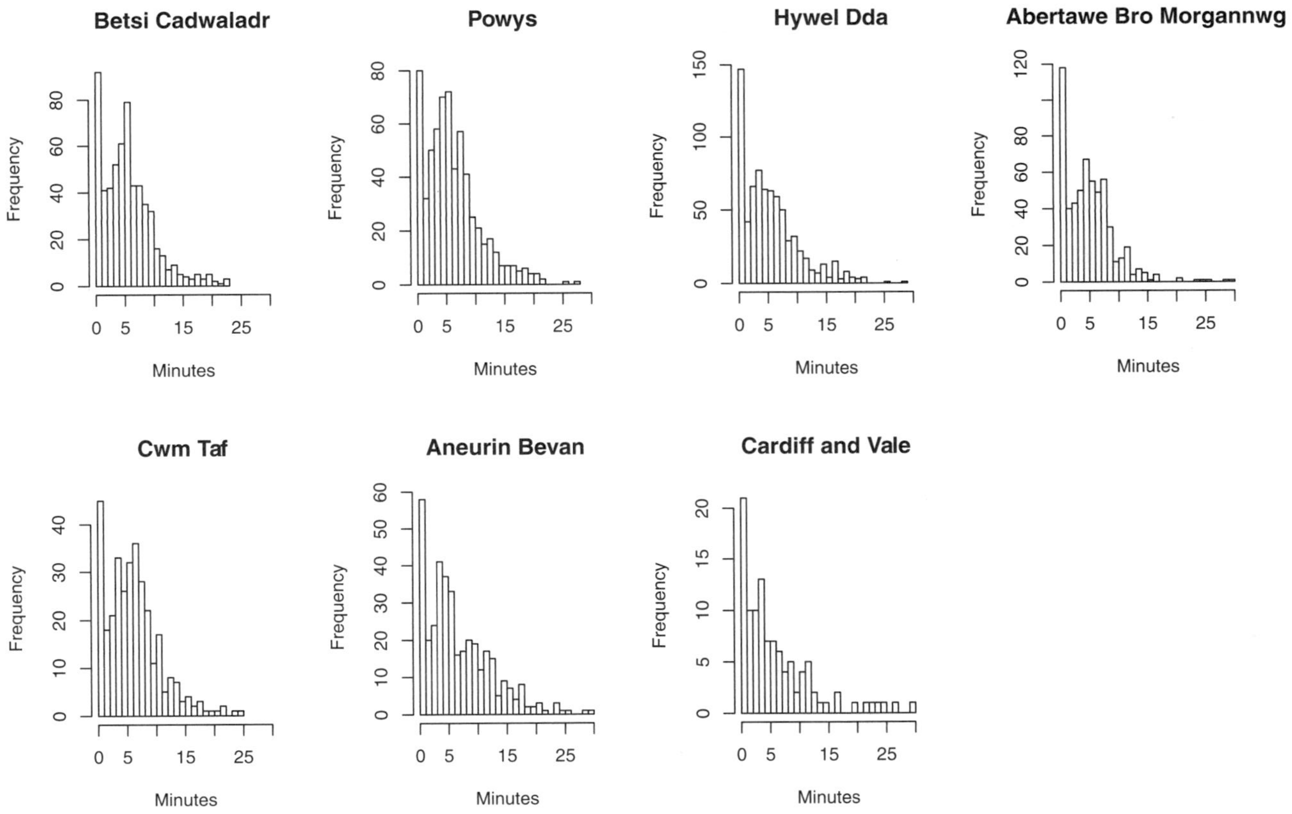

at its destination after the call has been received (i.e., ambulance travel time). We have selected the data for November 2015 and 2016 to compare across the 2 years. Selecting the same month from the 2 years is intended to eliminate seasonal effects. The travel time data are presented in 1-minute bins, but we will treat them as continuous data here. The bins are given their midpoint values, so the boundary is 0.5 minutes, and the longest travel times turn out to be 29.5 minutes. The histograms in Figure 2.3 for the seven regions all show a preponderance of cases on the boundary and another mode that is well away from the boundary.

We will compare two approaches to dealing with the boundary cases: hurdle models and Tweedie distributions. An advantage for the hurdle model is its ability to incorporate different effects for predictors of the boundary cases from those in the model for the nonboundary cases. A disadvantage is the larger number of parameters and therefore less parsimony (also the possibility of overfitting).

Starting with the hurdle model, the model for the zeros identifies an interaction between year and region due to the increased number of zeros for Abertawe Bro Morgannwg from 12% in 2015 to 28% in 2016. The main-effects gamma, Weibull, and lognormal models for the nonzero data are the best models, with no evidence of a year-by-region interaction. All three models identify 2016 as having shorter travel times than 2015 and Aneurin Bevan as requiring longer travel times than the other regions ($p < .0001$ for both effects).

The best Tweedie model is a main-effects model, the results of which are displayed in Table 2.5. It also indicates 2016 as having shorter travel

Table 2.5 Tweedie Model for Ambulance Travel Times

	Coefficient	Estimate	SE	t	p
	β_0	1.747	0.039		
Year (2016)	β_1	−0.209	0.030	−6.898	<.0001
Powys	β_2	0.089	0.049	1.802	.0717
Hywel Dda	β_3	−0.024	0.049	−0.493	.6217
Abertawe Bro Morgannwg	β_4	−0.130	0.053	−2.455	.0142
Cwm Taf	β_5	0.073	0.059	1.231	.2183
Aneurin Bevan	β_6	0.191	0.056	3.437	.0006
Cardiff Vale	β_7	0.074	0.091	0.819	.4131

Figure 2.4   Fitted Gamma, Tweedie, and Weibull Distributions

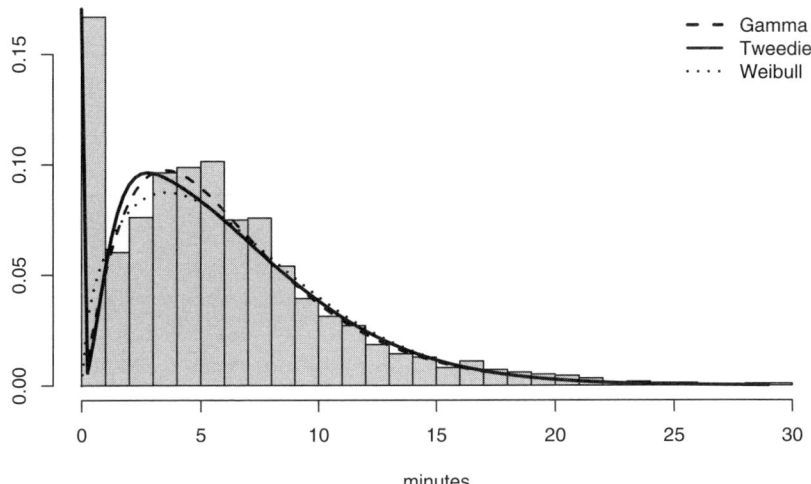

times than 2015 ($p < .0001$), Abertawe Bro Morgannwg as requiring shorter travel times ($p = .0142$), and Aneurin Bevan as requiring longer travel times ($p = .0006$).

Moreover, as Figure 2.4 illustrates (albeit for null models), the fitted distributions for the nonzero data in the hurdle model are similar to the Tweedie distribution over the nonzero domain. Given that the hurdle and Tweedie models present qualitatively similar findings, we may want to favor the Tweedie model on grounds of parsimony. The hurdle model has a total of 14 parameters, whereas the Tweedie model has just 8. The Tweedie model's AIC is 18,159.45, while the gamma, Weibull, and lognormal hurdle models' AICs are 18,151.04, 18,240.14, and 19,314.66, respectively. Thus, the Tweedie and gamma hurdle models are quite close competitors in terms of goodness of fit relative to parsimony.

The Wales health boards kindly provided relevant data on the characteristics of the seven regions, so we may investigate whether any of these characteristics account for the regional differences in ambulance travel time. In addition to population and area (and therefore population density), the available data include percentage of the population who are classified as disabled, percentage classified as carers, the percentage of those over 75 years of age living alone, the mortality rate for those under 75 years old, and the deprivation index. The best Tweedie model with these predictors retains the effect of the year and adds negative effects

Table 2.6   Tweedie Model with Regional Characteristics

	Coefficient	Estimate	SE	t	p
Intercept	β_0	2.137	0.093		
Year (2016)	β_1	−0.211	0.030	−6.960	<.0001
Population density	β_2	−0.0004	0.00008	−5.002	<.0001
Percent disabled	β_3	−0.076	0.024	−3.185	.0015

for population density and percentage classified as disabled. Both of these factors are associated with lesser ambulance travel times. Table 2.6 reports these effects. Unfortunately, the current software resources for Tweedie models do not include an option for a dispersion submodel, so we are unable to ascertain whether any of the regional characteristics affect variability in ambulance travel times. However, alternative hurdle models (not reported here) do not reveal any such effects.

CHAPTER 3. MODELS FOR DOUBLY BOUNDED VARIABLES

3.1 Doubly Bounded Variables and "Natural" Heteroscedasticity

In this chapter and the next chapter, we will introduce models for doubly bounded variables that are capable of modeling both location (e.g., the mean) and dispersion (e.g., variance). We begin by reviewing the challenges in modeling doubly bounded variables. We then introduce the beta distribution, currently the most popular for modeling doubly bounded variables, and describe regression models using this distribution. After that, we discuss the treatment of cases on the boundaries, via hurdle and mixture models (e.g., zero-inflation in proportional data).

As we indicated early in Chapter 1, there are two types of doubly bounded variables: those with true bounds and those with bounds due to censoring or truncation. Examples of the first type are proportions or rates (e.g., probability judgments, the proportion of recalled items in a memory test, the rate of change in income). This is the type of variable we are dealing with in this chapter and in Chapter 4.

Unlike variables without bounds, when variables are doubly bounded, the mean and the variance are no longer independent. As the mean approaches either bound, the variance must decrease. Likewise, the distribution becomes more skewed. Doubly bounded variables are naturally heteroscedastic and skewed, thereby making ordinary normal-distribution linear regression invalid. Some early solutions included modeling the variance as a function of linear or nonlinear combinations of predictor variables (e.g., Aitkin, 1987) and transforming variables to the real line. However, these solutions often are not satisfactory, and it has been a long-term challenge for modelers to construct GLMs for doubly bounded variables whose dispersion parameters (e.g., variances) are independent of location parameters (e.g., means). New advances in the field came when GLMs were introduced that employed distributions suitable for doubly bounded variables, such as the beta distribution.

3.2 The Beta Distribution: Definition and Properties

The beta distribution supports continuous variables within the $[0, 1]$ range and is well known for its flexibility in shapes. The beta distribution has two parameters. The parameterization we will use here has a mean, μ, and a "precision" parameter, ϕ, and we may describe a beta distribution as follows:

$$y \sim \text{Beta}\,(\mu\phi, \phi\,(1-\mu)) \in [0,1], \tag{3.1}$$

where $0 \le \mu \le 1$ and $\phi > 0$. The density function is

$$f\,(y|\phi,\,\mu) = \frac{\Gamma\,(\phi)}{\Gamma\,(\phi\mu) + \Gamma\,(\phi\,(1-\mu))} y^{\phi\mu-1}(1-y)^{\phi(1-\mu)-1}, \tag{3.2}$$

where $\Gamma\,(\bullet)$ denotes the gamma function.

Different values of μ and ϕ result in different distribution shapes such as skewed, unimodal, and U-shaped. When $\mu\phi$ and $\phi(1-\mu)$ are both equal to 1, we have the uniform distribution over the unit interval. If both $\mu\phi$ and $\phi(1-\mu)$ are greater than 1, the beta distribution has a single mode at $y = (\mu\phi - 1)/(\phi - 2)$. If both are less than 1, the distribution has a U-shape with an antimode at the same value of y. Finally, if $(\mu\phi - 1)(\phi(1-\mu) - 1) \le 0$, the distribution is J-shaped or reverse-J-shaped with no mode or antimode.

Figure 3.1 shows several shapes of the beta distribution with varying μ and ϕ. In this graph, for convenience, we have labeled the parameters

Figure 3.1   Example Beta Distributions

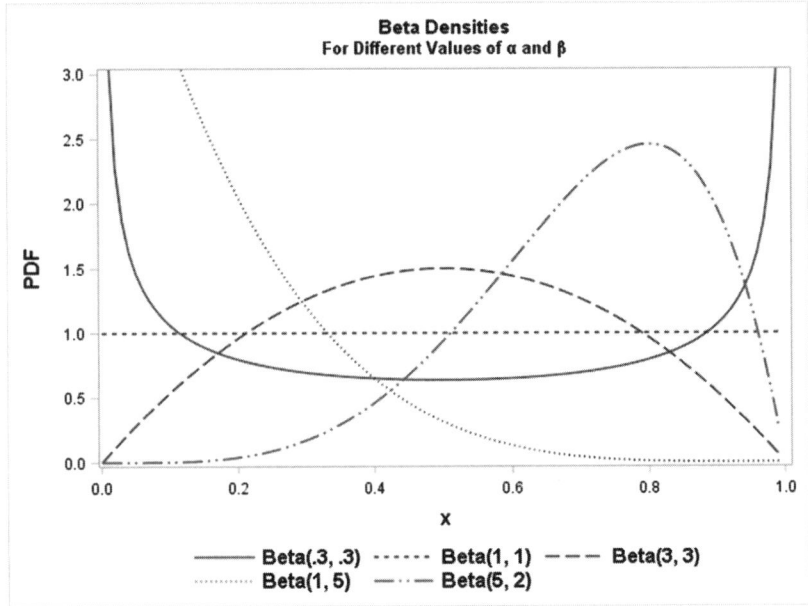

with $\mu\phi = \alpha$ and $\phi(1 - \mu) = \beta$, which is the standard way of parameterizing the beta distribution (this distribution and its properties are thoroughly described in Gupta & Nadarajah, 2004). Paolino (2001) and Ferrari and Cribari-Neto (2004) popularized the mean-precision parameterization.

The precision parameter, ϕ, covaries negatively with dispersion. We can see this from the variance, which is $\sigma^2 = \mu (1 - \mu) / (1 + \phi)$. As mentioned earlier, the variance and mean are not independent, but the precision parameter identifies the component of variance that does not depend on the mean. While ϕ is not the only parameter influencing the variance, it is unrestricted by the location parameter μ. Thus, we can separately model change in the location and change in the precision parameters. Smithson and Verkuilen (2006) provided a rationale for and demonstrations of this approach, and Verkuilen and Smithson (2012) extended it to mixed and multilevel models.

3.3   Modeling Location and Dispersion

3.3.1   *Location and Dispersion Submodels in the Beta GLM*

Let X and W be two covariate matrices associated with the location and precision of y, respectively. Their row vectors x_i and w_i are the ith independent observations, while column vectors X_k and W_k are the kth predictors. The linear combinations of X and W are $\eta_1 = X\beta$ and $\eta_2 = W\delta$, where β and δ are two column vectors of coefficients.

Let us assign a value of 1 to elements in X_0 and W_0, so that β_0 and δ_0 are two intercepts in the regression sense. Two link functions $f(\mu)$ and $h(\phi)$ are used to relate η_1 and η_2 with μ and ϕ of y. In the beta GLM, μ is in the interval (0, 1) and ϕ is positive, while η_1 and η_2 may not be. Therefore, $f(\mu)$ and $h(\phi)$ must satisfy the restrictions on y.

For the link between μ and η_1, candidate functions include logit, probit, cauchit, and log-log. The most commonly used is the logit link, which is defined by

$$f(\mu_i) = \eta_{1i} = \log(\frac{\mu_i}{1 - \mu_i}) = x_i\beta. \tag{3.3}$$

The advantage of using a logit link is that β can be interpreted as an odds ratio, as this allows for a straightforward interpretation of the importance of any β (Ferrari & Cribari-Neto, 2004). For detailed discussion of other candidate links, readers may refer to Collet (2003).

For the precision parameter ϕ, a log link conventionally is used to constrain η_2 to be positive:

$$h(\phi_i) = \log(\phi_i) = w_i\delta. \tag{3.4}$$

Because ϕ is a precision parameter, positive δ indicate a greater precision and thus a lower variance or dispersion. A negative sign could be added so that the model is an imprecision submodel, whereby an increase in δ also implies an increase in dispersion. Currently, programs such as betareg package (Cribari-Neto & Zeileis, 2010) endorse the precision submodel (modeling ϕ), while others have an alternative dispersion model. The gamlss package in R (Rigby & Stasinopoulos, 2005), for instance, uses $1/(1 + \phi)$.

Note that both the logit and log link functions require that the data do not include the boundary values 0 and 1. Readers will recall that log(0) is undefined. This limitation raises the question of how to deal with zeros and ones in the data, and we will deal with that in Section 3.5.

The location and dispersion submodels in beta regression are conditional on each other. The location submodel describing how predictors explain the mean of the dependent variable is defined by $f(\mu)$ with ϕ held constant. The dispersion submodel, which represents how predictors explain the precision parameter of the dependent variable, is defined by $h(\phi)$ with μ held constant.

An advantage of having submodels for location and dispersion is that modelers can evaluate separate influences on location and dispersion. For example, in psychopathology, variability in mood provides essential information about emotional stability. High variability in emotional states can be associated with mental disorders such as bipolar disorder. Events or circumstances that trigger the onset of a manic or a depressive state may not be the same as the situational factors influencing mood-swing amplitude itself.

3.3.2 Example 1: Measuring Speakers' Grammaticality Judgments

Lau, Clark, and Lappin (2014) investigated individual native speakers' judgments of the grammaticality of sentences generated via machine translation. They suggested that individual native speakers generally judge grammaticality along a continuum, rather than via a sharp binary acceptable versus unacceptable judgment. Their study used 2,500 sentences randomly selected from the British National Corpus. A random subset of 500 sentences was translated into one of the four target foreign languages (Norwegian, Spanish, Chinese, or Japanese) and then back-translated into English through machine translation.

These sentences were judged by participants on Amazon Mechanical Turk. They rated "naturalness" on three different types of measurement scales: a binary scale (accept vs. reject), a four-category scale, and a slider scale from 1 (*extremely unnatural*) to 100 (*extremely natural*). The researchers found that English sentences translated into Norwegian tended to yield the best results, and those translated into Japanese were the most unnatural. The researchers also found that for short to moderately long sentences, length had little influence on acceptability judgments.

We will focus on the slider scale data as these responses are close to being continuous data, and the scale is doubly bounded. There were a total of 247 sentences rated on the slider scale, including 49 original English sentences, 43 sentences back-translated from Spanish, 59 from Japanese, 47 from Norwegian, and 49 from Chinese. The sentence lengths ranged from 8 to 25 words. The number of independent ratings for each sentence ranged from 8 to 16. We used the average rating for each sentence (see a reanalysis of these data in Chapter 6 that takes interrater variability into account) as the dependent variable. Figure 3.2 shows the rating distributions for different language types.

Figure 3.2   Histograms for the Ratings in Grammar Data Example

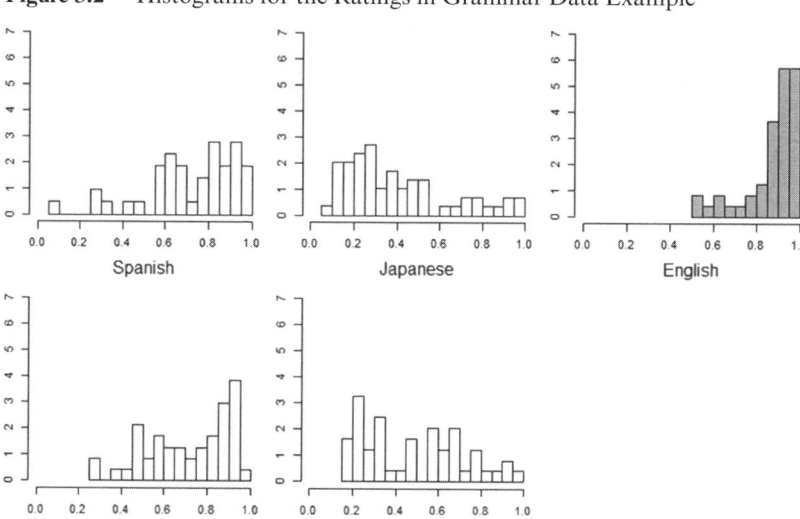

Figure 3.3   Association Between Ratings and Sentence Length

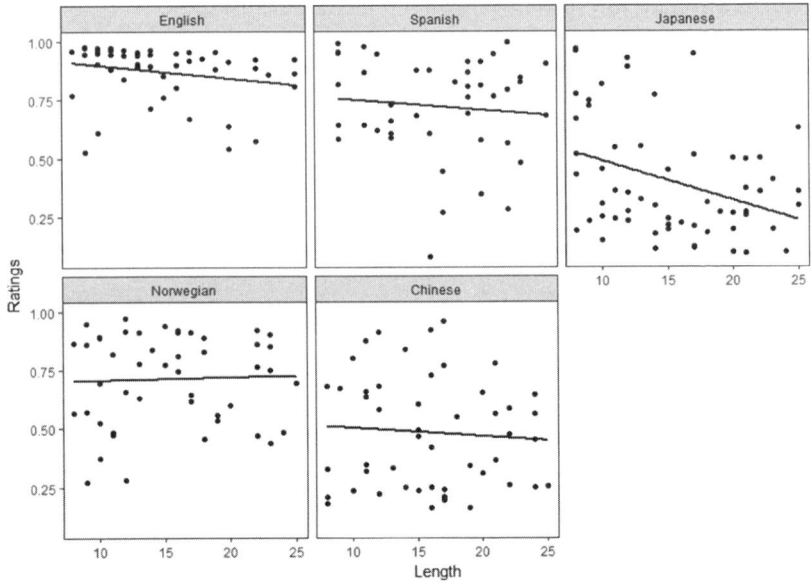

Ratings for sentences back-translated from Spanish and Norwegian are generally negatively skewed, while those from Japanese and Chinese are generally positively skewed and more spread out. Figure 3.3 shows the relationships between ratings and the sentence length across different languages. It is clear that the ratings of sentences translated from Japanese are strongly associated with sentence length: Longer sentences received lower ratings.

We analyzed the data using the beta GLM. Here we report the results using the betareg package in R, and results using SAS and Stata are available in the online supplementary materials. The ratings were divided by 100 to translate the data into the (0, 1) interval. Table 3.1 shows the model results for a model with language types and sentence length as predictors in location and precision submodel (the betareg package uses the precision model). The English original version is used as the reference group. The sentence length was centered by subtracting the mean length from the raw data.

The interaction between language types and sentence length was not significant in either the location submodel ($\chi^2 = 5.33$, $p = .255$, $df = 4$) or the precision submodel ($\chi^2 = 3.189$, $p = .527$, $df = 4$), so both

Table 3.1   Beta Regression Results for Grammaticality Example

Location Model					
Coefficient	Estimate	*SE*	z	2.5%	97.5%
(Intercept)	2.257	0.212	10.637	1.591	2.059
Spanish	−0.883	0.194	−4.541	−1.264	−0.502
Japanese	−2.190	0.170	−12.909	−2.523	−1.858
Norwegian	−0.897	0.174	−5.154	−1.238	−0.556
Chinese	−1.843	0.174	−10.612	−2.183	−1.502
Sentence length	−0.028	0.011	−2.547	−0.049	−0.006
Precision Model					
Coefficient	Estimate	*SE*	z	2.5%	97.5%
(Intercept)	1.791	0.315	5.692	2.002	2.802
Spanish	−1.169	0.287	−4.078	−1.731	−0.607
Japanese	−1.116	0.262	−4.262	−1.629	−0.603
Norwegian	−0.738	0.281	−2.623	−1.289	−0.187
Chinese	−1.051	0.273	−3.852	−1.586	−0.516
Sentence length	0.038	0.013	2.410	0.007	0.071

submodels contain only main effects for our predictors. We first inspect results in the location submodel. All other four languages received significantly negative coefficients, indicating that the ratings of sentences in all other four languages were substantially lower than the English original versions. And in line with the findings of the researchers, Japanese had the largest mean rating discrepancy from the English original ($b = -2.19$), followed by Chinese ($b = -1.84$). Spanish and Norwegian had similar discrepancies from the English original. Contrary to the original researchers' report, our model reveals a negative significant association between sentence length and ratings, indicating that longer sentences received lower rating in general.

There are two ways the location submodel coefficients can be interpreted: via mean differences or via odds ratios. Mean differences are useful for comparing states in categorical predictors. For example, how large are the differences in "naturalness" between the English- and Japanese-language sentences? The estimated mean rating for the English sentences is $\exp(2.257)/(1 + \exp(2.257)) = 0.905$, whereas for the Japanese sentences, it is $\exp(2.257 - 2.190)/(1 + \exp(2.257 - 2.190)) = 0.517$. Odds ratios are best suited for evaluating rates of change due to differences in values of continuous predictors or for comparing coefficients. For a one-word increase in sentence length, the predicted decline in "naturalness" on the odds scale is a factor of $\exp(-0.028) = 0.972$.

The results of the precision submodel shows that all other four languages received significantly negative coefficients. That implies that all the ratings for the four languages had lower precision, or larger conditional variance, than the ratings for the English original. Sentence length had a positive coefficient in the precision model, indicating that participants' ratings had smaller conditional variance with increasing sentence length.

The interpretation of precision submodel coefficients is a topic that currently needs further development, but here we present one way of doing so for categorical predictors. Recall that the variance in a beta distribution is influenced by both the mean and the precision. It makes sense to compare how variances would differ due to mean differences if the precision were constant versus how the variances differ when differences in precision also are taken into account. In our example, we have just seen that the predicted means for the English and Japanese sentences are 0.905 and 0.517, respectively. Their respective precisions are $\exp(1.791) = 5.995$ and $\exp(1.791 - 1.116) = 1.964$. As mentioned earlier, the variance is $\sigma^2 = \mu(1 - \mu)/(1 + \phi)$. So the estimated variance of the English sentences is $0.905(1 - 0.905)/(1 + 5.995) = 0.012$, and the variance of the Japanese sentences is $0.517(1 - 0.517)/(1 + 1.964) = 0.084$. The ratio of these variances is $0.084/0.012 = 6.855$. Now, if the English and Japanese sentences' precisions were identical, then the ratio of their variances would be $(0.517(1 - 0.517))/(0.905(1 - 0.905)) = 2.904$. So this variance ratio is $6.855/2.904 = 2.361$ times greater due to the differences in precision.

3.3.3 Example 2: Probability Judgment in Moral Dilemmas

We now turn to an example that involves an experimental design, and we also use it to address some issues regarding parameter estimation in beta GLMs. Shou and Song (2017) investigated moral decisions in variants of the trolley dilemma, in which people are required to choose between two options: killing an individual to save five imperiled people or not killing the individual but letting the five people die. The literature on judgments in moral dilemmas distinguishes between two types of moral dilemmas: personal and impersonal. Personal dilemmas involve personal action (e.g., pushing someone) when the killing takes place. Impersonal dilemmas involve the deflection of a threat. It has been commonly observed that people are more likely to prefer not killing in personal dilemmas than in impersonal dilemmas.

Shou and Song (2017) examined whether the difference in choice preferences between these two types of dilemma could be a result of the differences in perceived outcome probabilities between the two dilemmas. In their Experiment 1, they generated a personal dilemma called the "hostage dilemma" and an impersonal dilemma called the "car dilemma" (see Shou & Song, 2017, for details). Fifty-seven participants were presented with the car dilemma and 55 participants were presented with the hostage dilemma. The participants were asked to provide probability judgments for four types of outcomes:

1. $P(PO|K)$: probability that five people survive if killing is chosen

2. $P(NO|K)$: probability that one person dies if killing is chosen

3. $P(PO|\sim K)$: probability that five people die if not killing is chosen

4. $P(NO|\sim K)$: probability that one person survives if not killing is chosen

Figure 3.4 shows the rating distributions for the four outcome probabilities in the car and hostage dilemmas. Using independent-sample t tests, Shou and Song found that participants in the car dilemma provided significantly higher $P(PO|K)$ and $P(PO|\sim K)$ than did the participants in the hostage dilemma. On the other hand, participants' estimates for $P(NO|K)$ and $P(NO|\sim K)$ did not significantly differ between the two dilemmas.

We analyzed the data using the beta GLM and performed one beta regression model for each of the four outcome probabilities. Table 3.2 shows the results for the four types of probabilities. As indicated by the significantly negative coefficients in the location submodels, the mean $P(PO|K)$ and $P(PO|\sim K)$ in the hostage dilemmas were significantly lower than the car dilemma. Thus, participants had more pessimistic views regarding the outcomes of the hostage dilemma, indicating that the five people and the individual were less likely to survive in the hostage dilemma than in the car dilemma.

Turning now to the results in the precision submodel, the positive coefficient for $P(PO|K)$ suggests that the participants' judgments for this probability in the hostage dilemma had greater precision than those in the car dilemma. This could be primarily due to the fact that the responses in the car dilemma were highly skewed toward the right boundary (see Figure 3.4). The precision coefficient for $P(NO|\sim K)$ was also significantly positive, while the means of this probability did not

Figure 3.4   Probability Judgments Across Dilemmas

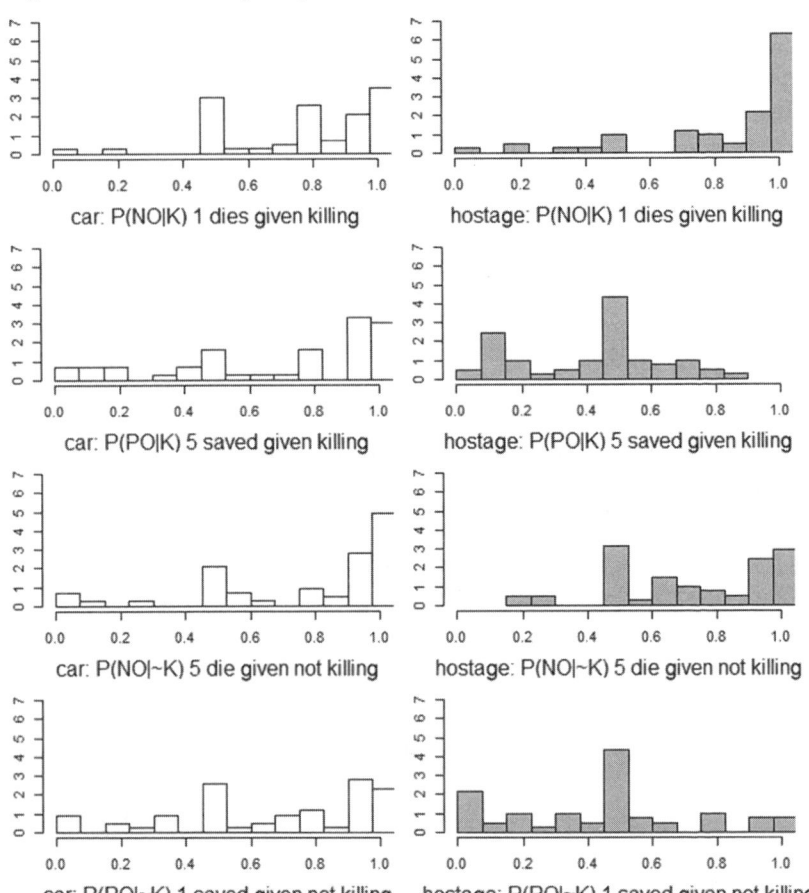

significantly differ between the two dilemmas. This suggests that the participants' judgments for $P(NO|{\sim}K)$ in the hostage dilemma might have less variance than those in the car dilemma.

3.4   Estimation and Model Diagnostics

3.4.1   *Maximum Likelihood Estimation and Estimator Bias*

Although the beta distribution is not a member of the exponential family (Ferrari & Cribari-Neto, 2004), maximum likelihood estimation

Table 3.2 Beta Regression Results for Moral Judgment Example

	Estimate	*SE*	*p*	Bias Corrected	
P(PO|K)				Coefficient	*SE*
Location—intercept	0.666	0.178	<.001	0.656	0.178
Hostage vs. car	−1.089	0.220	<.001	−1.077	0.221
Precision—intercept	0.023	0.157	.886	−0.011	0.156
Hostage vs. car	1.188	0.232	<.001	1.186	0.231
P(PO|~K)					
Location—intercept	0.587	0.172	.001	0.579	0.173
Hostage vs. car	−0.744	0.237	.002	−0.735	0.238
Precision—intercept	0.145	0.156	.352	0.112	0.155
Hostage vs. car	0.190	0.219	.387	0.190	0.219
P(NO|K)					
Location—intercept	1.135	0.182	<.001	1.121	0.183
Hostage vs. car	0.260	0.276	.346	0.252	0.277
Precision—intercept	0.420	0.177	.018	0.383	0.177
Hostage vs. car	−0.169	0.265	.523	−0.175	0.263
P(NO|~K)					
Location—intercept	1.023	0.192	<.001	1.008	0.192
Hostage vs. car	0.058	0.257	.820	0.063	0.259
Precision—intercept	0.004	0.172	.980	−0.033	0.171
Hostage vs. car	0.680	0.248	.006	0.679	0.247

(MLE) of the parameters in a beta GLM nevertheless generally is reliable. However, MLE in beta GLMs may incur two problems. The first one is similar to many other GLMs, namely, that MLE can be biased when the sample size is small, as the order of approximation to the asymptotic distribution is $O(1/n)$ and the asymptotic standard error has order $O(1/\sqrt{n})$. The second issue relates to estimating the precision parameter. The standard error of the precision parameter estimate usually is relatively large and has downward bias in comparison to the location parameter (Verkuilen & Smithson, 2012; Noel, 2014). The two issues aggravate each other, because estimation bias in the precision parameter can be more pronounced when the sample size is small (Noel, 2014).

Several approaches have been proposed to address MLE bias in beta GLMs, either via a correction after obtaining the estimates (bias correction) or a preventive correction (bias reduction) prior to obtaining the estimates. In early bias correction approaches, the bias-corrected MLEs were not suitable for regression models (Cordeiro, Da Rocha, Da Rocha, & Cribari-Neto, 1997; Ospina, Cribari-Neto, & Vasconcellos, 2006). Simas, Barreto-Souza, & Rocha (2010) and Ospina et al. (2006) proposed bias-corrected MLEs for beta GLMs.

Let us denote the MLEs of the beta GLM parameter vectors by $\hat{\beta}$ and $\hat{\delta}$. The bias-corrected estimators are $\tilde{\beta} = \hat{\beta} - \hat{\mathbf{B}}\left(\hat{\beta}\right)$ and $\tilde{\delta} = \hat{\delta} - \hat{\mathbf{B}}\left(\hat{\delta}\right)$, where $\hat{\mathbf{B}}\left(\hat{\beta}\right)$ and $\hat{\mathbf{B}}\left(\hat{\delta}\right)$ are the MLEs of the second-order bias of $\hat{\beta}$ and $\hat{\delta}$. The alternative approach, bias reduction, focuses on modifying the score function prior to obtaining the parameter estimates. The modified component scores for parameters proposed by Firth (1993) are $U^{*}(\beta) = U(\beta) + K(\beta)B(\beta)$, and $U^{*}(\delta) = U(\delta) + K(\delta)B(\delta)$.

$K(\bullet)$ is the information matrix, and $B(\bullet)$ represents the $O(1/n)$ bias. Bias adjustment may also be done by resampling methods such as bootstrapping. Each pseudo-sample gives an estimator β^{*} and δ^{*}. $\hat{\beta^{*}}$ and $\hat{\delta^{*}}$ are the expected values of the distributions of β^{*} and δ^{*} obtained from a large number of pseudo-samples. The bias value based on the bootstrap estimators is the difference between the $\hat{\beta^{*}}/\hat{\delta^{*}}$ and $\hat{\beta}/\hat{\delta}$. The corrected estimators via bootstrap are $\tilde{\beta} = \hat{\beta} - (\hat{\beta^{*}} - \hat{\beta}) = 2\hat{\beta} - \hat{\beta^{*}}$ and $\tilde{\delta} = \hat{\delta} - (\hat{\delta^{*}} - \hat{\delta}) = 2\hat{\delta} - \hat{\delta^{*}}$.

In Table 3.2, we show bias-corrected parameter estimates for both the location and precision submodels. We can see that in our example, bias correction has made little difference to the estimates, although there is a slight tendency for the bias corrections to produce less extreme values. In our experience, bias correction usually does not result in dramatic differences, although the aforementioned authors do present examples where the differences are substantial.

3.4.2 Model Fit, Comparison, and Effect Sizes

For model comparisons, the likelihood ratio test statistic (i.e., twice the difference between the log-likelihoods of the two models) usually can be used. Likewise, commonly used model fit statistics such as the AIC and BIC can be used. In the grammaticality example, we used the likelihood ratio test to determine the significance of the interaction terms in the model. The log-likelihood for a model with the main effects of language and sentence length in both submodels is 100.354 ($df = 12$). Adding an interaction effect between language and sentence length in the location submodel increases the log-likelihood to 103.022 ($df = 16$), while adding an interaction effect in the precision submodel results in a log-likelihood value at 101.949 ($df = 16$). The differences can be tested using the χ^2 tests, which suggest that both interaction effects are not significant ($\chi^2 = 5.336, p = .255, df = 4$ for the location interaction, and $\chi^2 = 3.189, p = .527, df = 4$ for the precision submodel).

Turning now to effect size measures, we already have observed that location-parameter coefficients can be interpreted either by differences in predicted means or by using odds ratios in much the same way as is done for logistic regression models. We observed that mean differences are easier to interpret and are natural for comparisons between states in categorical predictors (e.g., between experimental conditions). Odds ratios are more useful as an effect size measure for continuous predictors or for comparing coefficients in the location submodel.

A corresponding effect size measure for the precision submodel is not readily available, but we suggested such a measure in our first example. Recall that we compared the ratio of variances in sentence quality for English versus Japanese, under the assumption that the precisions were identical, with the ratio of the variances given the actual precision estimates. We found that the variance ratio was 2.361 times greater due to the differences in precision between the two languages. More precisely, the 2.361 factor was due to the ratio $(\phi_j + 1)/(\phi_e + 1)$, where ϕ_j denotes the precision for the Japanese sentences and ϕ_e denotes the precision for the English sentences. So, we recommend using this ratio, $(\phi_1 + 1)/(\phi_2 + 1)$, as an effect size measure for precisions.

We will try this on the second example. Recall that the participants' judgments for $P(NO|\sim K)$ in the hostage dilemma had less variance than those in the car dilemma. From Table 3.2, the predicted mean for the hostage dilemma is $\mu_h = \exp(1.023)/(1 + \exp(1.023)) = 0.736$, and the mean for the car dilemma is $\mu_c = \exp(1.023 + 0.058)/(1 + \exp(1.023 + 0.058)) = 0.747$. The difference between these means is not significant, and if the precisions in the two dilemmas were identical, then the ratio of the two variances would be close to 1: $\sigma_c^2/\sigma_h^2 = \mu_c(1 - \mu_c)/(\mu_h(1 - \mu_h)) = 0.972$. However, the estimated precisions in the two dilemmas are $\phi_h = \exp(0.004) = 1.004$ and $\phi_c = \exp(0.004 + 0.680) = 1.982$. The ratio $(\phi_c + 1)/(\phi_h + 1) = 1.488$, so the estimated variance ratio is $0.972 * 1.488 = 1.487$.

What about an overall effect size measure that takes into account effects on means and precisions? Borrowing again from the logistic regression literature, consider the proportional reduction of error

(PRE) statistics or pseudo-R^2, which imitate the linear regression R^2 by assessing a models goodness of fit relative to the null model. Most of the popular PRE measures in logistic regression, such as McFadden's pseudo-R^2 and Nagelkerke's pseudo-R^2, assume that the likelihood falls between 0 and 1 (Long, 1997). This assumption does not hold in beta GLMs because the likelihood of an individual observation in beta distributions is based on the density and can be greater than 1. One candidate that can be used in beta GLMs is Cox and Snell's (1989) measure:

$$PRE = 1 - \left(\frac{L_{null}}{L_M} \right)^{2/N}. \tag{3.5}$$

L_{null} is the likelihood of the null model, L_M is the alternative model, and N is the sample size. Cox and Snell's PRE can be regarded as a transformation of the likelihood ratio statistic and interpreted as the geometric mean squared improvement in individual observations (Menard, 2000).

By now, readers may be wondering what statistics they should report for a beta regression. At this time, there are no fixed standards and specific requirements may vary across disciplines. That said, we would recommend reporting at least the coefficients and standard errors for both the location and precision submodels. Confidence intervals for these also would be advisable (as in Table 3.1). For interpretation, either the means difference or odds ratio summaries for the location submodel and the precisions ratio summary for the precision submodel can be used, although it is very likely that these will need to be explained to reviewers and editors. Where model comparison is performed, the appropriate test and statistics should be reported (e.g., log-likelihoods and likelihood ratio chi-square statistics).

3.4.3   Residuals and Influence Statistics

Influential cases may be diagnosed by residuals. However, readers should note that currently there is no consensus on the best residual measures of beta GLMs. Raw residuals used in other GLMs (i.e., the differences between observed responses and predicted/fitted responses) provide limited and even distorted information due to natural heteroscedasticity and the boundedness of the dependent variable in beta GLMs.

Early proposed residuals include Pearson and deviance residuals. The Pearson residuals simply scale the raw residuals using the square root of variance function:

$$r_i = \frac{y_i - \hat{\mu}_i}{\sqrt{\hat{\mu}_i \left(1 - \hat{\mu}_i\right) / \left(1 + \hat{\phi}_i\right)}}. \tag{3.6}$$

The deviance residuals apply the scaled log-likelihood of the model:

$$r_i = sign\left(y_i - \hat{\mu}_i\right) \sqrt{2\ell_i \left(y_i, \hat{\phi}_i\right) - 2\ell_i \left(\hat{\mu}_i, \hat{\phi}_i\right)}. \tag{3.7}$$

Three weighted versions of residuals are proposed by Espinheira, Ferrari, and Cribari-Neto (2008) and are based on the difference between the logit of y_i and its MLE expected value under an adopted model. The first version is defined as

$$r_i = \frac{y_i^* - \hat{\mu}_i^*}{\sqrt{\hat{\phi}_i v_i}}, \tag{3.8}$$

where $\hat{\mu}_i^*$ is the expected value of y_i^*, and v_i equals the variance of y_i^*.

A standardized version of this residual, also called sandwich residual Type 1, is given by

$$r_i = \frac{y_i^* - \hat{\mu}_i^*}{\sqrt{v_i}}. \tag{3.9}$$

Using an alternative way to standardize this residual, a second type of sandwich residual can be obtained:

$$r_i = \frac{y_i^* - \hat{\mu}_i^*}{\sqrt{v_i \left(1 - h_{ii}\right)}}, \tag{3.10}$$

where h_{ii} is the ith diagonal item of the hat matrix H.

Espinheira et al. (2008) argued that the weighted versions of residuals could be more powerful for identifying model misspecification. The sandwich Type 2 residual could account for observation leverages and is better approximated by the standard normal distribution than the raw or Pearson residual. The three weighted types of residuals are also more likely to identify the atypical observations that strongly influence estimates of regression parameters.

Figure 3.5 illustrates the different types of residuals using the language rating example. Notice that Cases 70 and 77 were singled out by the three weighted residuals. Both cases are ratings of Spanish back-translated sentences. Removing the two cases changed the location

Figure 3.5   Residuals for Grammar Data Example

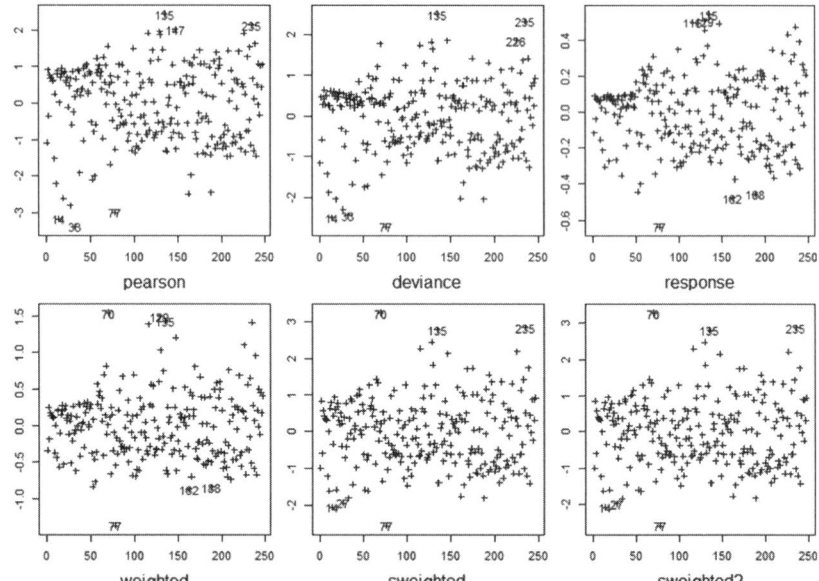

submodel coefficient for Spanish from –0.88 to –0.82 and the precision submodel coefficient for Spanish from –1.17 to –0.88. Thus, including the two cases substantially increases the dispersion of the ratings in the Spanish-language group, biasing the coefficient estimate.

Influential cases also may be identified via Cook's distance and dfbeta parameters, both of which are based on casewise deletion methods to investigate the influence of each individual case. Cook's distance assesses a scaled change in fitted values when the ith observation is deleted. The dfbeta assesses the change in each parameter estimate when the ith observation is removed and scaled by the standard error of the estimated parameter.

3.5   Treatment of Cases at the Boundaries

3.5.1   Rescaling from [0, 1] to (0, 1): Methods and Issues

Because the link functions in beta regression are undefined at 0 and 1, the regression model can be fitted only to data in the (0, 1) interval. However, variables such as probabilities and rates have the true boundaries at 0 and 1, and these values may occur in the data. When there are only a small number of zeros or ones, a practical remedy is to linearly transform the boundary values into the open interval. A popular transformation is

$$y'' = \frac{y'(N-1) + c}{N},$$ (3.11)

where N is the sample size and c is a compression parameter that "pushes" the boundary values away from the boundary. A typical value for c is 0.5.

The smaller the value of c is, the closer the boundary values are to 0 or 1, but they also have greater impact on estimation of the precision parameters. When there are many boundary values in the data, transforming them away from the boundary is no longer justifiable or practical. A typical example is a ceiling effect in credit card debt payments, where most cardholders pay off 100% of their debt in every payment period. Hurdle models may be used to deal with zeros and ones, which we introduce in the next subsection.

3.5.2   Hurdle and Boundary-Inflated Models

A general class of hurdle models has been proposed by Ospina and Ferrari (2012). These models often are referred to as "zero-inflated" or "ones-inflated" models, but they are in fact hurdle models as they aim at separating the boundary values from the rest of the data and apply a discrete-continuous two-stage modeling process. The discrete part fits a Bernoulli or a degenerate distribution at 0 or 1, while the continuous part fits a beta distribution to the data in the (0, 1) interval.

The mixed density for a hurdle model dealing with zeros is

$$f(y; \pi_0, \mu, \phi) = \begin{cases} \pi_0, & \text{if } y = 0 \\ (1 - \pi_0) f(y; \mu, \phi), & \text{if } 0 < y < 1, \end{cases}$$ (3.12)

and its mean and variance are

$$E(y) = (1 - \pi_0)\mu,$$

$$Var(y) = (1 - \pi_0)\frac{\mu(1-\mu)}{\phi + 1} + \pi_0(1 - \pi_0)\mu^2.$$

For a model dealing with ones, the density is

$$f(y; \pi_1, \mu, \phi) = \begin{cases} (1 - \pi_1) f(y; \mu, \phi), & \text{if } 0 < y < 1 \\ \pi_1, & \text{if } y = 1, \end{cases}$$ (3.13)

and its mean and variance are

$$E(y) = \pi_1 + (1 - \pi_1)\mu,$$

$$Var(y) = (1 - \pi_1)\frac{\mu(1 - \mu)}{\phi + 1} + \pi_1(1 - \pi_1)(1 - \mu)^2.$$

Ospina and Ferrari (2012) introduced a general class of zero-/one-hurdle beta regression by including a third submodel in addition to the location and precision submodels. The third submodel models the discrete component and evaluates the likelihood of a data point falling in the boundary component versus the (0, 1) component:

$$f(\pi_i) = \eta_{3i} = \log(\frac{\pi_i}{1 - \pi_i}) = Z_i\theta, \tag{3.14}$$

where Z is the covariate matrix associating with the discrete component of y, and θ is the vector of parameters. The two are combined in η_3 and are linked with the π via a logit link. The mixture distribution can be expanded to have all three components: zero, one, and the rest in the interval (0, 1). Further extensions and model details are elaborated in Ospina and Ferrari (2012).

To illustrate the use of a hurdle model, we reanalyze the moral judgment data from our second example. As noted in Figure 3.4, for the positive outcomes of the two options, there were more participants rating 100% for the car dilemma than for the hostage dilemma. We apply a ones-hurdle beta GLM. The results are displayed in Table 3.3.

Table 3.3   One-Inflated Hurdle Model for the Moral Judgment Example

	Coefficient	SE	2.5%	97.5%
Discrete (One) Component				
Intercept	−1.674	0.257	0.084	0.570
Hostage vs. car	−1.900	0.639	−1.056	−0.429
Continuous (Beta) Component				
Location submodel				
Intercept	0.327	0.122	0.110	0.574
Hostage vs. car	−0.742	0.157	0.285	0.944
Precision submodel				
Intercept	0.342	0.157	−2.180	−1.171
Hostage vs. car	0.614	0.232	−3.153	−0.647

The discrete component shows a significant negative coefficient for the comparison between the hostage dilemma and the car dilemma. This indicates that the responses in the hostage dilemma were less likely to be at 1 than those in the car dilemma: $\exp(-1.674)/(1 + \exp(-1.674)) = 0.158$ versus $\exp(-1.674 - 1.900)/(1 + \exp(-1.674 - 1.900)) = 0.027$, or an odds ratio of $\exp(1.900) = 6.686$. For the rest of the values, participants still provide lower subjective probabilities for positive outcomes in the hostage dilemma than in the car dilemma.

CHAPTER 4. QUANTILE MODELS FOR BOUNDED VARIABLES

4.1 Introduction

A viable strategy for analyzing bounded data is quantile regression. Hao and Naiman (2007) outline three major limitations of linear regression models: (1) providing limited information about the behavior of a dependent variable for values far from the conditional mean, (2) inability to deal with heteroscedasticity, and (3) providing limited information about the distribution shape, such as skewness or bimodality. A quantile regression can estimate any quantile such as the 25th percentile, median, or the 75th percentile. Researchers using this technique often prefer to model several quantiles, because together they can reveal whether the distribution of the dependent variable changes shape for different values of a predictor (e.g., whether the skew covaries with a predictor). Moreover, quantile regression does not require the strong distributional assumptions that we need in linear regression.

In this chapter, we will introduce quantile regression and provide an example with a bounded dependent variable. However, we will also argue for a different modeling strategy, namely, the cumulative density function (CDF)-quantile GLM. We have published articles about this new approach (Smithson & Shou, 2017), and in the second half of this chapter, we will explain why the CDF-quantile model may have advantages over the basic quantile model. We also will present an example of where it has advantages over the beta regression method described in Chapter 3.

4.2 Quantile Regression

Early work on conditional median models was extended to other quantiles of the distribution by Koenker and Bassett (1978), who elaborated the foundations of quantile regression. Quantile regression is akin to a traditional GLM of a location parameter, whereby a specific quantile is predicted by a linear combination of independent variables. One can select several quantiles and construct a regression model for each quantile, to differentiate among the effects of covariates on different parts of the distribution. In this section, we will introduce quantile regression and provide an example.

Suppose we have a linear regression model,

$$y'_i = \beta_0 + \boldsymbol{\beta}\boldsymbol{x_i} + \epsilon_i, \tag{4.1}$$

where y'_i is the predicted value of the dependent variable Y, the right-hand side is a combination of regression coefficients ($\boldsymbol{\beta}$) and predictor variables $\boldsymbol{x_i}$, and ϵ_i is the residual. Extending this to a quantile regression will require some additional terms and notation. Recall that the tth quantile of Y is the score below which $t * 100\%$ of the values in Y fall. We denote the tth quantile of Y with a parenthetical superscript: $y_i^{(t)}$. We will use this superscript for the other terms in a quantile regression model. Thus, the tth conditional quantile given $\boldsymbol{x_i}$ is

$$y'^{(t)}_i = \beta_0^{(t)} + \boldsymbol{\beta}^{(t)}\boldsymbol{x_i} + \epsilon_i^{(t)}. \tag{4.2}$$

It may not be immediately clear why a different set of parameters is required for every quantile, so let us consider a toy example. Suppose we have just a binary predictor, $x = \{0, 1\}$, and we want to use it to predict the median (the 50th quantile) and the other quartiles (25th and 75th quantiles). Now suppose that when $x = 0$, the 25th, 50th, and 75th quantiles of Y are 0.3, 0.4, and 0.8, and when $x = 1$, the 25th, 50th, and 75th quantiles of Y are 0.2, 0.6, and 0.85, respectively. Then the predicted conditional quantiles in our quantile regression models are

$$\begin{aligned}
y'^{(0.25)}_i &= \beta_0^{(0.25)} + \beta_1^{(0.25)}x_i = 0.3 - 0.1x_i \\
y'^{(0.5)}_i &= \beta_0^{(0.5)} + \beta_1^{(0.5)}x_i = 0.4 + 0.2x_i \\
y'^{(0.75)}_i &= \beta_0^{(0.75)} + \beta_1^{(0.75)}x_i = 0.8 + 0.05x_i.
\end{aligned} \tag{4.3}$$

Similar to the ordinary least squares (OLS) linear regression, parameter estimation in quantile regression can be achieved by minimizing residuals, but instead of the sum of squared residuals, quantile regression minimizes the sum of absolute residuals. To simplify notation, let us denote $\beta_0^{(t)} + \boldsymbol{\beta}^{(t)}\boldsymbol{x_i}$ by $\xi^{(t)}$, so the residuals are $y_i^{(t)} - \xi^{(t)}$. The coefficients for the tth quantile regression are estimated by minimizing the weighted sum of absolute residuals:

$$\min_{\xi \in R} \sum_{i=1}^{n} \rho_t |y_i^{(t)} - \xi^{(t)}|. \tag{4.4}$$

The ρ_t weights are determined by the quantile rank (i.e., proportions t and $1 - t$):

$$\sum_{i=1}^{n} \rho_t |y_i^{(t)} - \xi^{(t)}| = \sum_{i; y_i \geq \xi} t|y_i^{(t)} - \xi^{(t)}| + \sum_{i; y_i < \xi} (1 - t)|y_i^{(t)} - \xi^{(t)}|. \quad (4.5)$$

The median estimator $y_i^{(0.5)}$ is a special case because $t = 1 - t = 0.5$, so the absolute residuals are equally weighted and therefore only their sum needs to be minimized: $\min_{\xi \in R} \sum_{i=1}^{n} |y_i - \xi|$.

Quantile regression allows for the assessments of covariates' effects over different quantile locations in the distribution of the dependent variable, thereby enabling researchers to examine effects of heteroscedasticity and skew. In our toy example, we saw that as x goes from 0 to 1, the 25th quantile decreases but the 75th quantile increases, thereby telling us that the dependent variable's dispersion increases as x goes from 0 to 1. Quantile regression parameter estimates are also less sensitive to outliers and extreme values than their counterparts in linear regression models. Overall, then, quantile regression is able to provide information about effects on distribution shape, spread, and skewness that elude linear regression (see the next section for an example).

Another advantageous property of quantile regression is *monotonic equivariance*. Sometimes researchers apply monotonic transformations such as the log to correct nonnormality in the dependent variable. Monotonic equivariance ensures that the conditional quantiles of the transformed dependent variable are identical to the transformed conditional quantiles of the original dependent variable. For example, the quantiles of $\log(y)$ are the log of quantiles of y. Thus, interpretations of the quantile regression results still hold when the dependent variables are transformed back into the original scale. Monotonic equivariance does not generally hold for conditional means modeled in linear regression, as many nonlinear transformations do not scale each value in the data equally. So, this is an advantage that quantile regression has over linear regression.

Because of monotonic equivariance, it becomes feasible to interpret quantile regression effects and reconstruct the effects of the covariates when the dependent variable has been monotonically transformed. Hao and Naiman (2007) described two types of effects to assist the interpretation of the effect of predictor variables on quantiles of the dependent variables. One is called typical-setting effect, which is interpreting the change of one predictor variable on the various quantiles of the dependent variable while holding all other predictor variables fixed at a typical value (such as the mean). The second is to use the mean effect, where the effect of a predictor variable on a quantile of the dependent variable is averaged over all individual cases in the data.

Quantile regression does have one potential pitfall that researchers must keep in mind, namely, that it can produce nonunique solutions. Most software packages running quantile regression will alert users to nonunique solutions when they occur.

4.2.1   Example: Depressive Symptoms among Chinese University Students

Yu et al. (2015) conducted a large-scale study that involved 6,000 university students in China. The aim of the study was to investigate the influence of family environment on the level of depression among the university students. The family environment is measured by the Family Environment Scale that assesses nine relational traits (see Table 4.1 for details about the nine traits). The level of depression was measured by the Beck Depression Inventory (BDI; 21 items; Beck, Steer, & Brown, 1996).

Figure 4.1 shows the histogram of the BDI scores. The BDI scores are positively skewed, because the sample is a nonclinical sample. The authors conducted traditional linear regression and ANOVA to examine the associations between the family environment factors and various demographic information (e.g., maternal education levels and parental relationship). The authors found that several family factors such as cohesion, conflict, control, and organization had significant associations with BDI scores. The mean of BDI scores also was influenced by the aforementioned demographic variables.

We use this data set to illustrate quantile regression. We select five quantile locations: 0.1, 0.25, 0.5, 0.75, and 0.9. Our model has five predictor variables, including three family relational traits (cohesion, conflict, and control) and two demographic variables (maternal education and parental relationship). Figure 4.2 presents the coefficients of different predictors on the different quantiles of the BDI scores. In each panel, the black dots identify the slope coefficient of a predictor for the quantile numbered on the x-axis. The horizontal solid and dashed lines indicate the coefficient and its confidence interval of a linear regression. The figure clearly shows that the coefficients of the various predictors in lower and upper quantiles are very different from the linear regression

Table 4.1 Family Environment Scale Relational Traits

Family Traits
Cohesion: the level of commitment, help, and support among the family members.
Expressiveness: the amount to which family members are encouraged to express their feelings.
Conflict: the extent of open expression of anger and conflict between family members.
Independence: the degree of esteem, self-confidence, and independence among family members.
Achievement orientation: the extent to which family members view general activities (such as study or work) as achievement oriented or competitive.
Intellectual: cultural orientation, the degree of interest in political, intellectual, and cultural activities.
Active-recreational orientation: the extent to which family members join recreational activities.
Moral-religious orientation: the degree of emphasis on ethnicity, religion, and value in family members.
Organization: organization and structure for planning family activities and assigning responsibilities.
Control: the extent to which family members use the rules and procedures to arrange their life.

coefficients. For example, the coefficient for control increases with the quantile of the BDI, suggesting that the association between control and depression increases among students who are more depressed.

We can also use the results in quantile regression to examine the change in distribution shape. One way to detect the change in the distribution is to inspect the change in the intercept. As all the predictor values are centered in our model, the value of intercept represents the fitted quantile function when all predictor variables are at their respective means. As shown in Figure 4.2, the slope line of the intercept becomes steeper as the quantile increases. The slope above the median (e.g., the line from the 75th quantile to the 90th quantile) is steeper than the slope below the median (e.g., the line from the 10th quantile to the 25th quantile). This also means that the predicted change in the dependent variable from the 10th quantile to the 25th percentile is smaller than it is from the

Figure 4.1   The Histogram of the BDI Scores

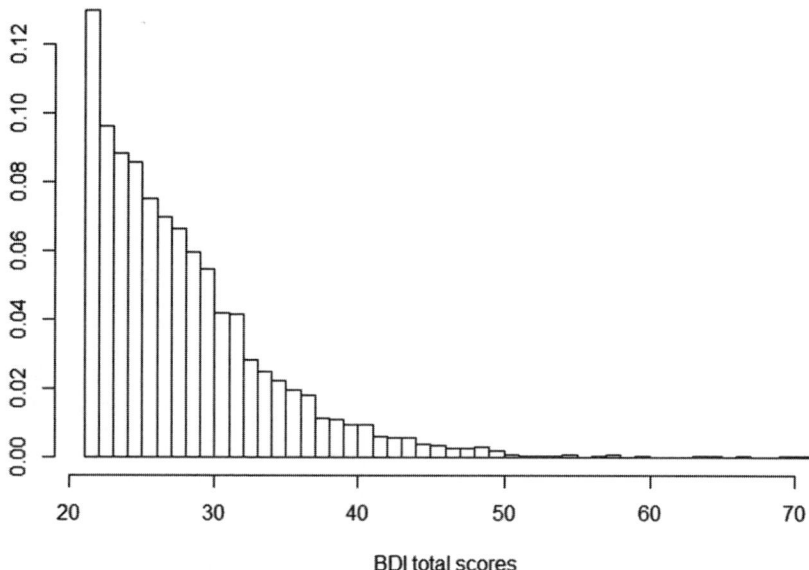

BDI total scores

75th quantile to the 90th quantile, despite the same interquantile range (0.15). This is typical when the dependent variable is positively skewed.

This pattern can also be observed at the quantitative level by computing the predicted interquantile range changes. As explained in Hao and Naiman (2007), one can compare the coefficient of a covariate or intercept at the tth quantile ($\beta^{(t)}$) and the coefficient for the same covariate at the $(1 - t)$th quantile ($\beta^{(1-t)}$). If the distribution is symmetric and there is no change in dispersion, there should be no difference between $\beta^{(1-t)}$ and $\beta^{(t)}$. For the BDI scores, the intercept is 21.64 at the 10th quantile and 34.51 at the 90th quantile. These clearly differ, indicating asymmetry and/or change in dispersion.

This approach can also be used to detect if a predictor variable influences the scale shifts. In our example, the coefficients of conflict at the 10th, 25th, 75th, and 90th quantiles are 0.12, 0.21, 0.45, and 1.05, respectively. The interquantile range changes resulting from the conflict trait are 0.09 for the 10th versus 25th quantiles and 0.6 for the 75th versus 90th quantiles. This suggests that the effect of the conflict trait variable covaries with the BDI scores.

Figure 4.2   Estimated Coefficients of Predictor Variables across Different Quantiles of the BDI

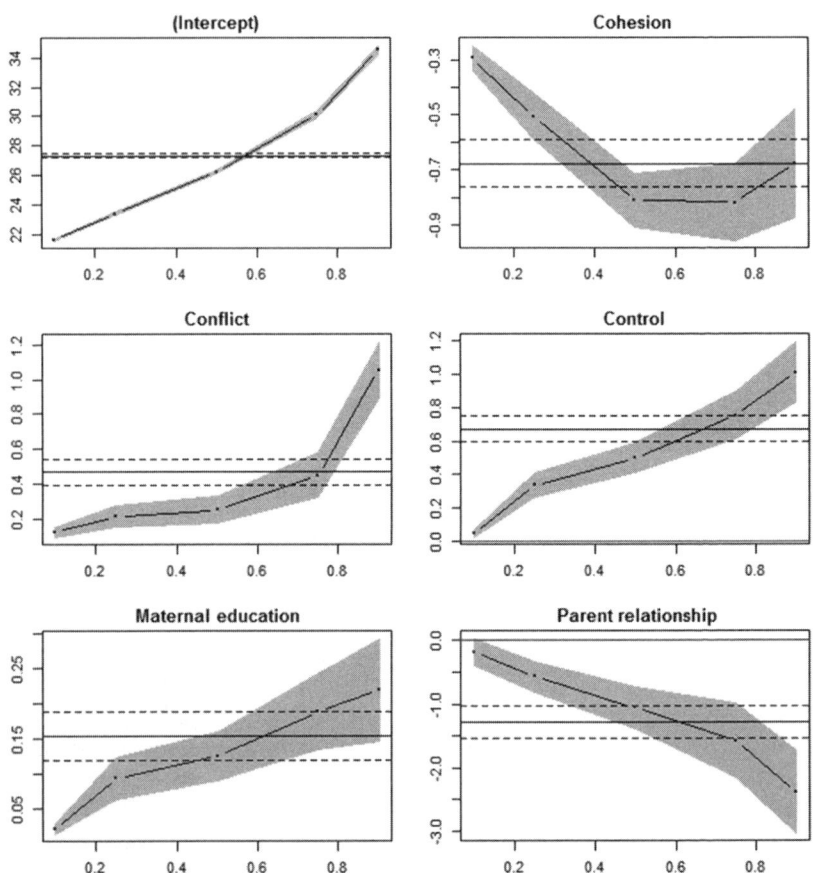

NOTE: The *y*-axis is the coefficient value, and the *x*-axis is the quantiles of the BDI. The horizontal lines are the ordinary least squares coefficients (solid) and their confidence bounds (dash). The gray areas are the 95% confidence polygon around the quantile coefficients.

4.3   Distributions for Doubly Bounded Variables with Explicit Quantile Functions

Quantile regression is an attractive alternative to linear regression that can deal with skew and heteroscedasticity. However, each quantile requires its own set of parameter estimates, and there are infinitely many quantiles. Thus, quantile regression cannot describe all quantiles with a finite number of parameters. Moreover, quantile regression is based on the assumption that the dependent variable (at each quantile) is from a distribution with unbounded support. This assumption does not hold for doubly bounded random variables. An alternative quantile-based approach is a class of GLMs based on two-parameter distributions whose parameters determine the conditional location and spread of all quantiles. The next subsection introduces the family of distributions on which these GLMs are based.

4.3.1 The CDF-Quantile Family

Popular methods for analyzing doubly bounded random variables involved transforming them onto the entire real line, thereby effectively eliminating the bounds. The earliest of these, Johnson's SB (N. L. Johnson, Kotz, & Balakrishnan, 1995) distribution, transforms a random variable on the (0, 1) interval to a normal distribution via the logit transformation, that is, transforming a doubly bounded random variable from (0, 1) to the real line and fitting a normal distribution to the transformed variable. Tadakimalla and Johnson (1982) introduce a logit-logistic distribution by replacing the normal distribution in Johnson's SB distribution with the standard logistic distribution.

Transformation to the real line results in tractable models but sacrifices the ability to directly interpret effects on location and dispersion in the (0, 1) interval setting, as beta regression does. Smithson and Shou (2017) extend the transformation approach by applying the cdf of one standard distribution to the quantile function of another standard distribution. Their procedure transforms a doubly bounded random variable from (0, 1) to either the real line or nonnegative half of the real line and then back to the (0, 1) interval. Location and dispersion parameters are estimated in "mid-transformation", prior to the final transformation back to the (0, 1) interval. They name the resulting distribution family the CDF-quantile distributions.

The members of this family have explicit probability density functions (pdfs), cumulative density functions (cdfs), and quantile functions. They are amenable to likelihood inference and thereby both maximum likelihood and Bayesian estimation techniques. Unlike the beta distribution, they enable a wide variety of quantile regression models for random variables on the (0, 1) interval with predictors for both location and dispersion parameters and simple interpretations of those parameters. Furthermore, the family can model a wide variety of distribution shapes, with greater skew and kurtosis coverage than the beta. Some members of this family share special cases and behavior at the boundary of the (0, 1) interval with the beta distribution.

Let $G(x, \mu, \sigma)$ denote the cdf of a CDF-quantile distribution with support (0, 1), a real-valued location parameter μ, and positive scale parameter σ. We define G as

$$G(x, \mu, \sigma) = F[U[R[H^{-1}(x)], \mu, \sigma]], \qquad (4.6)$$

where F is a standard cdf with support, which we shall denote by D_1; H is an invertible standard cdf with support denoted by D_2; $R : D_2 \to D_1$ is an invertible monotonic function; and $U : D_1 \to D_2$ is an appropriate transform for imposing the location and scale parameters. We limit the domains D_1 and D_2 to pairs taken from $[-\infty, \infty]$ and/or $[0, \infty]$, and consider two special cases of U. For $D_1 = [-\infty, \infty]$, we define

$$U(y, \mu, \sigma) = (y - \mu)/\sigma, \qquad (4.7)$$

and for $D_1 = [0, \infty]$, we have

$$U(y, \mu, \sigma) = (e^{-\mu} y)^{1/\sigma}, \qquad (4.8)$$

where $y = R[H^{-1}(x)]$. So, $H^{-1}(x)$ transforms x from (0, 1) to D_2 (e.g., the real line). Then μ acts as a location parameter by relocating x, and σ acts as a dispersion (or scale) parameter by rescaling x. Finally, F takes the real-valued $U(x, \mu, \sigma)$ and returns its cdf value in the F distribution, which is in the (0, 1) interval.

When F and H exchange roles, we have "quantile-duals" of one another in the sense that one's cdf is the other's quantile, with the appropriate parameterization. These distributions' names take the form of "F-H" (e.g., Cauchit-Logistic and Logit-Cauchy). If F is invertible, then for $D_1 = [-\infty, \infty]$ and $G(x, \mu, \sigma) = \gamma$, the quantile is

$$G^{-1}(\gamma, \mu, \sigma) = H[R^{-1}[(F^{-1}(\gamma) + \mu/\sigma)\sigma]], \qquad (4.9)$$

and for $D_1 = [0, \infty]$, it is

$$G^{-1}(\gamma, \mu, \sigma) = H[R^{-1}[(e^{\mu/\sigma} F^{-1}(\gamma))^{\sigma}]]. \qquad (4.10)$$

We now elaborate the logit-logistic distribution as an example with support pair $D_1 = D_2 = [-\infty, \infty]$. This distribution employs the logit function $\frac{1}{1+e^{-z}}$ for both F and H. Inverting H and applying it and F to equation 4.6 gives

$$G(x,\mu,\sigma) = \cfrac{1}{1 + e^{\frac{\mu}{\sigma}}\left(\frac{x}{1-x}\right)^{\frac{1}{\sigma}}}, \qquad (4.11)$$

and differentiating it gives the pdf

$$g(x,\mu,\sigma) = \cfrac{1}{-\sigma(x-1)^2\left[e^{-\frac{\mu}{\sigma}}\left(\frac{x}{1-x}\right)^{\frac{1}{\sigma}} + e^{\frac{\mu}{\sigma}}\left(\frac{x}{1-x}\right)^{\frac{1}{\sigma}} + 2\right]}. \qquad (4.12)$$

Inverting equation 4.11, we obtain the quantile function:

$$G^{-1}(\gamma,\mu,\sigma) = \cfrac{1}{1 + \exp\left(-\left(\frac{\mu}{\sigma} - \log(\frac{1-\gamma}{\gamma})\right)\sigma\right)}. \qquad (4.13)$$

Note that this equation immediately tells us that the median is a function solely of μ: $G^{-1}(1/2,\mu,\sigma) = 1/(1 + \exp(-\mu))$.

Smithson and Shou (2017) introduce a total of 36 self-dual distributions for which the parent distribution supports are $D_1 = D_2 = [-\infty,\infty]$ (see Table 4.2 for the list of 36 distributions). They show that for all distributions in this family, the median is a function solely of μ, and σ controls the spread of the other quantiles around the median. Thus, μ is a location parameter and σ is a dispersion parameter. The CDF-quantile

Table 4.2 Distributions with Support $D_1 = D_2 = [-\infty,\infty]$ by Shape

			H			
F	Logistic	Cauchy	ArcSinh	T2	Burr VII	BurrVIII
Logit	LL*	BM	BM	BM	LL*	BM
Cauchit	TM	FT*	FT	TM	TM	TM
ArcSinh	TM	FT	FT*	TM	TM	TM
T2	TM	BM	BM	FT*	TM	TM
Burr VII	LL*	BM	BM	BM	LL*	BM
Burr VIII	FT	BM	BM	BM	FT	FT*

Note: LL = logit-logistic shape; BM = bimodal shape; FT = finite-tailed shape; TM = trimodal shape. *Includes the uniform distribution as a special case.

Figure 4.3 Four Subfamilies of CDF-Quantile Distributions

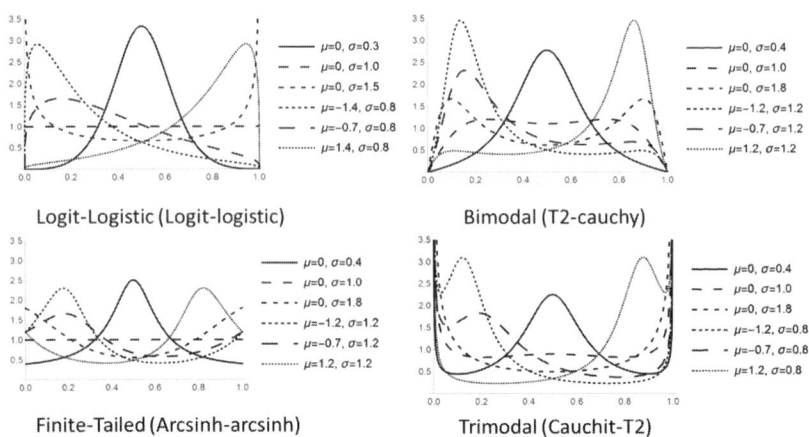

Logit-Logistic (Logit-logistic)   Bimodal (T2-cauchy)

Finite-Tailed (Arcsinh-arcsinh)   Trimodal (Cauchit-T2)

distributions also are mirror-inverted by changing the sign of μ. They are symmetric when $\mu = 0$. Equations 4.9 and 4.10 provide an expression for every quantile with just two parameters. Therefore, if we have predictors for μ and/or σ, then that provides a model of all quantiles with a finite number of parameters, unlike quantile regression.

These distributions are classified into four groups as they capture four kinds of characteristic shapes: logit-logistic shape (LL), bimodal shape (BM), trimodal shape (TM), and finite-tailed (FT) shape. The LL subgroup has a shape typified by the logit-logistic distribution and shares some properties with the beta distribution, such as including the uniform as a special case, but covers different combinations of skewness and kurtosis. The BM subgroup is able to capture bimodality on the (0, 1) interval and does not include the uniform distribution as a special case but has symmetric special cases at $x = 0.5$. The third group, the FT group, can capture heavy tails at the boundaries of the (0, 1) interval. Finally, the TM group has two of its modes at 0 and 1. Figure 4.3 shows the examples for the four groups of distributions.

4.4   The CDF-Quantile GLM

The CDF-quantile family of distributions can be used to form generalized linear models (GLMs). Similar to the beta GLMs introduced in Chapter 3, the CDF-quantile GLMs have two submodels that link the linear combinations of predictors with the two parameters μ and σ: $L_\mu(\hat{\mu}) = \mathbf{x}^T \beta$, and $L_\sigma(\hat{\sigma}) = \mathbf{z}^T \delta$, where $\mathbf{x}$ and $\mathbf{z}$ are vectors of predictors (the sets of predictors may or may not overlap) and β and δ are vectors of coefficients. For location submodel of μ, the identity link can be used for distributions whose supports are $D_1 = D_2 = [-\infty, \infty]$, and the log link can be used for distributions whose supports are $D_1 = D_2 = [0, \infty]$. For the dispersion submodel of σ, the log link can be used for either pair of supports.

The location submodel also provides a model for the median, as the quantile function reduces to a function of μ alone. From equation 4.9, if F is invertible, then for $D_1 = [-\infty, \infty]$ and some real number $c \neq -\mu$, we have $F^{-1}(1/2) = c/\sigma$. The median is then $Q_{.5} = H(\mu + c)$, which is a function of μ. For support $D_1 = [0, \infty]$, from equation 4.10, we have $F^{-1}(1/2) = b^{c/\sigma}$, where $b^{c/\sigma} \neq e^{-\mu/\sigma}$. In this case, the median is $Q_{.5} = H(e^\mu b^c)$. For example, the logit-logistic distribution's quantile function yields $F^{-1}(1/2) = 0$, and therefore, as mentioned earlier, $Q_{.5} = G^{-1}(1/2, \mu, \sigma) = 1/(1 + \exp(-\mu))$.

Because σ is a dispersion parameter, positive δ coefficients in $\mathbf{z}^T \delta$ indicate a positive association between the predictor variables and the dispersion of the dependent variable. As mentioned earlier, σ controls the spread of quantiles around the median, so greater dispersion implies that other quantiles will be further away from the median.

We now turn to maximum likelihood estimation. For $D_1 = D_2 = [-\infty, \infty]$, the pdf can be written as

$$g(x, \mu, \sigma) = \frac{q(x) f\left(\frac{H^{-1}(x) - \mu}{\sigma}\right)}{\sigma}, \tag{4.14}$$

where f is the pdf corresponding to F, and q is the quantile distribution function corresponding to H^{-1}. Smithson and Shou (2017) show that as f is differentiable and twice differentiable, it is possible to obtain explicit expressions for the gradient and Hessian, from which standard errors of the parameter estimates may be obtained. The distributions in Table 4.2 are amenable to likelihood inference and thereby both maximum likelihood and Bayesian estimation techniques. Fortunately, the nonuniqueness solution problem that can occur in quantile regression does not apply to CDF-quantile GLMs. The reason for this is that whereas a quantile regression model estimates just one quantile, CDF-quantile models estimate all quantiles with just the conditional μ and σ parameters.

4.4.1 Example

We use the Yu et al. (2015) data from the example in Section 4.1.1 to illustrate the application of a CDF-quantile GLM and to compare with the quantile regression approach. The Beck Depression Inventory (BDI) scores are doubly bounded due to the BDI scale limits, and here for illustrative purposes, we are treating the bounds as absolute. The BDI scores are transformed into the $[0, 1]$ interval via the linear equation: $y = (y - a)/(b - a)$, where a is the lowest possible value of BDI score and b is the highest possible value. Next, the range is compressed into the $(0, 1)$ interval by taking $y'' = (y'(N - 1) + c)/N$, where N is the sample size and c is the compressing parameter.

The histogram in Figure 4.1 shows a unimodal skewed distribution shape. So we choose the logit-logistic distribution for modeling this dependent variable. Details of the model fitting can be found in the online supplementary material for this chapter. The five predictors in Figure 4.2—cohesion, conflict, control, maternal education, and parental relationship—are included in both the location and dispersion submodels. Table 4.3 shows the model estimation results.

Table 4.3 Logit-Logistic CDF Quantile Regression for Predicting BDI Scores

Location model					
Coefficient	Estimate	*SE*	z	2.5%	97.5%
(Intercept)	−2.541	0.030	−85.127	−2.600	−2.480
Cohesion	−0.193	0.015	−13.296	−0.222	−0.165
Conflict	0.114	0.013	8.675	0.088	0.140
Control	0.184	0.014	12.992	0.156	0.211
Maternal education	0.038	0.006	6.124	0.026	0.050
Parent relationship	−0.255	0.041	−6.118	−0.337	−0.174
Dispersion model					
Coefficient	Estimate	*SE*	z	2.5%	97.5%
(Intercept)	0.107	0.013	8.182	0.082	0.133
Cohesion	0.132	0.008	15.649	0.115	0.148
Conflict	−0.045	0.007	−6.239	−0.059	−0.031
Control	−0.031	0.008	−3.754	−0.047	−0.015
Maternal education	−0.014	0.003	−4.207	−0.021	−0.008
Parent relationship	0.093	0.024	3.827	0.046	0.142

All five predictors have significant effects in both location and dispersion submodels. The directions of effects of the five predictors in the location submodel are the same as those found by quantile regression. Family cohesion and harmony of parental relationship have negative effects on the median of the BDI score, while family conflict, control, and maternal education have positive effects.

Turning to the results in the dispersion submodel, again the coefficients of the five predictors are significant. As mentioned in the previous section, positive coefficients in the dispersion model indicate a positive association between predictors and the dispersion (σ) of the dependent variable. As shown in Table 4.3, the cohesion has a positive coefficient in the dispersion model, indicating that scoring higher on family cohesion is associated with a greater value in σ of the BDI scores. This result is related to the results in the location submodel that cohesion has a negative relationship with BDI.

4.4.1.1 *Example Comparing CDF-Quantile with Beta Regression*

We apply the logit-logistic distribution to modeling the grammar study data in Chapter 3 and compare results with the beta regression model in that chapter. This type of comparison is limited by the fact that beta regression and CDF-quantile regression address different questions about the data. We shall elaborate points of comparison between these two models, because these will provide insights into how the CDF-quantile and beta regression models differ not only in the nature of their parameters but also in what they are predicting.

To begin, a beta regression estimates a conditional mean and conditional precision and thereby indirectly estimates a conditional variance. Because the variance of a beta distribution has a $\mu(1-\mu)$ term, any location submodel effect pushing the mean toward 1/2 increases variance, and pushing the mean away from 1/2 decreases variance. The precision parameter, then, tracks effects on the variance that are not accounted for by effects on the mean. The predicted values in a beta regression model, then, are obtained in the usual way for a general linear model (i.e., via an equation for the conditional expected value connecting it with the weighted linear combination of predictors through the link function).

A CDF-quantile model, on the other hand, estimates conditional quantiles. The μ parameter tracks the median, and the σ parameter tracks the spread of the other quantiles around the median. Consequently, the CDF-quantile and beta regression models may not always agree in their portrayals of conditional central tendency. Smithson and

Table 4.4 Logit-Logistic CDF Quantile Regression for Predicting Grammar
Ratings

Location model					
Coefficient	Estimate	*SE*	z	2.5%	97.5%
(Intercept)	2.261	0.124	18.165	2.017	2.505
Spanish	−0.959	0.236	−4.066	−1.421	−0.497
Japanese	−2.860	0.192	−14.926	−3.236	−2.485
Norwegian	−1.083	0.207	−5.221	−1.489	−0.676
Chinese	−2.327	0.207	−11.250	−2.732	−1.921
Sentence length	−0.042	0.014	−2.681	−0.069	−0.016
Dispersion model					
Coefficient	Estimate	*SE*	z	2.5%	97.5%
(Intercept)	−0.707	0.121	−5.847	−0.944	−0.470
Spanish	0.471	0.177	2.661	0.124	0.819
Japanese	0.291	0.163	1.788	−0.028	0.609
Norwegian	0.266	0.170	1.570	−0.066	0.599
Chinese	0.308	0.169	1.824	−0.023	0.638
Sentence length	−0.029	0.011	−2.681	−0.050	−0.008

Shou (2017) present an example where these two models make opposite
predictions for the means versus medians.

Table 4.4 shows the logit-logistic model estimation results for our
example. In this case, the location submodels for the logit-logistic and
beta models have significant effects in the same directions (see Table 3.1).
All four non-English languages receive significantly negative coefficients,
indicating that the ratings of these languages were substantially lower
than for the English counterparts. Likewise, longer sentences receive
lower ratings, as indicated by the negative coefficient for length. These
results suggest that the conclusions we may draw about central tendency
are fairly robust, because they hold for conditional medians as well as
for conditional means.

On the other hand, the dispersion submodels differ between the two
models. In the logit-logistic regression, all four non-English languages
receive positive coefficients, but only the coefficient for Spanish is significant. In contrast, all coefficients in the beta regression precision
submodel are significant (and negative, indicating lower precision and
therefore greater dispersion). Given that means and precisions are more

Table 4.5 Logit-Logistic Model Estimates and Predictions

	μ	σ	Estimate 50th	Actual 50th	Estimate 25th	Actual 25th	Estimate 75th	Actual 75th
English	2.261	0.493	0.906	0.915	0.848	0.851	0.943	0.951
Spanish	1.303	0.790	0.786	0.765	0.607	0.604	0.898	0.873
Japanese	−0.599	0.660	0.355	0.311	0.210	0.217	0.531	0.515
Norwegian	1.179	0.644	0.765	0.765	0.616	0.561	0.868	0.887
Chinese	−0.065	0.671	0.484	0.466	0.310	0.250	0.662	0.658

sensitive to outliers than quantiles, the divergent conclusions of these two models suggest that we might want to investigate whether they could be due to the influence of outliers on the beta regression results.

Given that a CDF-quantile model estimates conditional quantiles, with the μ parameter identifying the median and the σ parameter controlling the spread of the other quantiles around the median, a reasonable question to raise is how well the model tracks various quantiles. Table 4.5 displays the conditional μ and σ estimates for each of the languages and compares the estimated and actual 50th, 25th, and 75th conditional quantiles. For instance, from Table 4.4, the Chinese $\mu = 2.261 − 2.327 = −0.065$, and $\sigma = \exp(−0.707 + 0.308) = 0.671$. Thus, the Table 4.5 entries are the marginal effects of language in the location and dispersion submodels.

The quantile function gives us the quantiles predicted via marginal effects. In our example, the logit-logistic quantile function is

$$y = 1/\left(1 + \exp\left(−\mu + \sigma \log\left((1 − \gamma)/\gamma\right)\right)\right), \qquad (4.15)$$

where γ is the quantile rank. For instance, the English-language estimated median is $y = 1/\left(1 + \exp\left(−2.261 + 0.493 \log\left((1 − .5)/.5\right)\right)\right) = 1/(1 + \exp(−2.261)) = 0.906$, and the estimated 25th percentile is $y = 1/\left(1 + \exp\left(−2.261 + 0.493 \log\left((1 − .25)/.25\right)\right)\right) = 0.848$. These estimates are fairly close to their counterparts in the data (0.915 and 0.851, respectively).

However, the model does not fit all of the languages equally well. English and Spanish are best fitted, with average absolute errors of 0.007 and 0.016, respectively. Japanese, Norwegian, and Chinese have worse fits, with average absolute errors of 0.022, 0.025, and 0.027, respectively. These are due mainly to worse fits for the 25th and 50th quantiles.

70

Figure 4.4   Raw, Pearson's, and Deviance Residuals in BDI Data Example

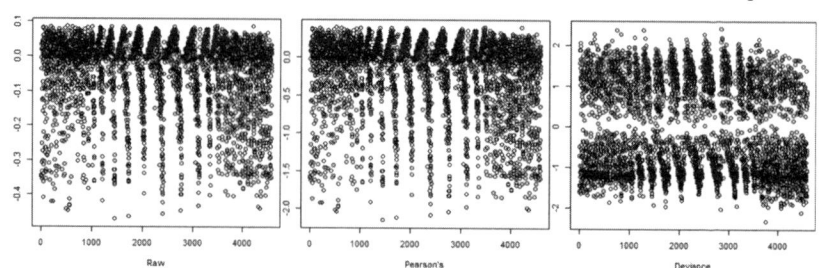

4.4.2   Residuals and Influence Measures

Commonly used residuals such as the Pearson, $r_i = (y_i - \hat{\mu}_i)/\hat{\sigma}_i$, and deviance residuals can be applied to inspect the model residuals in CDF-quantile GLMs. For the data example above, the raw, Pearson's, and deviance residuals are shown in Figure 4.4. As is the case for beta regression, residuals for CDF-quantile GLMs constitute an active research topic. Little is known about the behavior of residuals for these models, and there also is the consideration that the parameters of these distributions are directly related to quantiles rather than to distribution moments.

The relative influence of individual observations can be assessed via dfbetas. A dfbeta value for an observation is the scaled difference between a parameter value estimated in a model by using the entire sample and a parameter value estimated in the model using the sample with that observation deleted.

4.4.3   Guidelines, Unresolved Problems, and Extensions

Currently, there is no systematic guide regarding the selection of CDF-quantile distributions for empirical model fits. The subfamilies' shape features can provide some ad hoc guidance for distribution selection by referring to empirical distribution shapes. Researchers may wish to test several candidate distributions. One advantage of this distribution family is that researchers can assess the robustness of their findings by ascertaining whether models using alternative CDF-quantile distributions yield similar results.

Similar to beta GLMs, the CDF-quantile GLMs have limitations regarding the treatment of boundary values. Only members of the FT subfamily have density defined at 0 and 1. If there are not many cases at the bounds, the linear transformation defined in Chapter 3 can be applied to shift boundary cases away from 0 and 1. As with beta regression, a smaller value of c will leave such cases closer to the boundaries, and they will have a greater impact on the estimation of the conditional dispersion parameter. When there are many boundary values, a hurdle model can be constructed that applies the two-step estimation procedure described in Chapter 3. Thus, a logistic regression can model the probability that a case will be at the boundary (either at 1 or 0) as the first step, and a CDF-quantile GLM can then model the rest of the distribution.

CHAPTER 5. CENSORED AND TRUNCATED VARIABLES

We define and discuss varieties of truncation and censoring in this chapter, including fixed versus random threshold models and sample selection models. We elaborate Tobit models because they are the most well known, and we focus on the most common type of Tobit model with brief introductions to the more complex and less commonplace types. Model estimation, evaluation, and diagnostics are covered. Our main example is a double-censoring model (i.e., where both the lower and upper bounds of a scale are censored scores). After discussing issues of nonnormality and heteroscedasticity for Tobit models, we extend this example by fitting a non-Gaussian heteroscedastic model.

5.1 Types of Censoring and Truncation

We introduced censoring and truncation in Chapter 1, where we distinguished among different kinds of bounded variables. The crucial distinction between censoring and truncation is that cases that are censored on one or more variables nevertheless have other variables whose values are observed, whereas cases that are truncated are omitted from the sample altogether so that none of their variables are observed. Thus, truncation is purely a matter of sample selection. Censoring may or may not involve sample selection, and if it does, then it is limited to the variables in which censoring occurs.

This distinction is illustrated in Figure 5.1 with an example of lower-bound censoring and truncation. In this example, a bank's losses are modeled with a lognormal distribution. However, the loss amounts are recorded only if they exceed $1,000. In the censored distribution, the losses with censored values are recorded as $1,000, which generates a spike at the threshold $\tau = 1,000$. The truncated distribution simply excludes these cases from the distribution. The lower-truncated sample's distribution function is rescaled to have an area of 1 by dividing it by $1 - F(\tau)$. This normalized distribution function is represented in the right-hand graph by the dashed-line curve outlining the lighter-gray area.

As mentioned in Chapter 1, censoring may arise in two ways. First, it may occur when boundaries are an artifact of the measuring instrument.

Figure 5.1   Censored versus Truncated Distributions

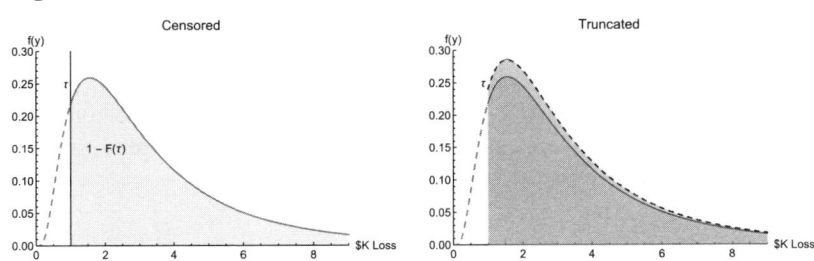

These include scales whose scores are the sums of responses on items whose response formats have bounds (as in the agree-disagree response formats typical of attitude surveys) and scales whose bounds are not anchored at absolute minima or maxima (e.g., a scale measuring a person's income with an upper bound "more than X").

Second, censoring may arise from a sample selection procedure, in a similar fashion to truncation. Sample selection may be decided by a criterion built into the measurement process itself, as when the amount of time spent dwelling on a webpage has an upper bound determined by a time-out mechanism in the website. If dwell time is a measure of the amount of interest a person has in the webpage material, then the timed-out cases' degrees of interest are censored observations. Alternatively, censoring may be determined by another random variable. For instance, we may wish to model household mortgage debts in which the sample selection variable is whether the home is mortgaged or not. This sample selection variable, in turn, has its own expectation and sampling error, and its model includes predictors such as household income that also are predictors for the mortgage debt model.

Finally, the censoring thresholds may be fixed or unknown with a threshold-determining random process. A brief description of the latter kind of model is provided in the next section. These sometimes are referred to as "Type I" and "Type II" censoring, respectively, but these are not to be confused with the same terminology popularized by Amemiya (1984) for distinguishing Tobit models that involve sample selection from those that do not (these also are described in the next section).

Modeling censored or truncated data both involves latent variable models. There are three kinds of estimates from the models for y_i that may interest researchers: those based on the latent variable (y', say), those based on the conditional truncated variable (e.g., $y|y' > \tau$ in the lower-truncation case), and, for censoring models, those based on the censored variable y itself. Interest in these will depend on the researcher's goals, but the most commonly employed estimators are those for the latent variable. As for sample selection models of τ_i, these may resemble hurdle models if the censoring involves a sharp or qualitative distinction between censored and uncensored cases or boundary-inflated models when τ_i and y_i share predictors.

5.2 Tobit Models

We begin by reprising the lower-censored Tobit model introduced in Chapter 1. As before, the traditional notation describes the uncensored observations as

$$y_i = x_i\beta + e_i, \tag{5.1}$$

where $\epsilon_i \sim N(0, \sigma)$. For the variable's censored observations, if τ is the censoring threshold,

$$x_i\beta + e_i \leq \tau, \tag{5.2}$$

so that $e_i \leq \tau - x_i\beta$ and therefore a model for the censoring rate is

$$\Pr(y_i \leq \tau \,|\, x_i) = 1 - \Phi((x_i\beta - \tau)/\sigma), \tag{5.3}$$

where Φ is the standard normal cdf.

A latent variable version of this model yields Amemiya's (1984) Type I Tobit model:

$$y_i = \begin{array}{l} y'_i \text{ if } y'_i > \tau \\ \tau \text{ if } y'_i \leq \tau \end{array} \tag{5.4}$$

where y' denotes the latent variable, with $y'_i = x_i\beta + e_i$. Assuming a single distribution for Y, the likelihood function for the censored regression model is

$$L_i = \prod_{y_i=\tau} F(\tau) \prod_{y_i>\tau} f(y_i), \tag{5.5}$$

where F and f are the cdf and pdf of the distribution of Y, respectively. The likelihood function for the truncated regression model is

$$L_i = \prod_{y_i>\tau} f(y_i)/(1 - F(\tau)). \tag{5.6}$$

Equations (5.5) and (5.6) are fixed-threshold models. Random-threshold models require replacing τ with τ_i in each of them. These models then have two submodels, for y_i and τ_i, which may or may not share predictors. Models of this kind are closely related to survival analysis, a broad class of techniques including non- and semiparametric as well as parametric models. Survival analysis is beyond the scope of this book, and its literature includes book-length treatments (e.g., Tableman & Kim, 2004).

A doubly censored regression model is a straightforward extension of the model developed here, and we will examine an example of it later in this chapter. We have two thresholds, τ_1 and τ_2, so that

$$y_i = \begin{array}{l} y'_i \text{ if } \tau_2 > y'_i > \tau_1 \\ \tau_1 \text{ if } y'_i \leq \tau_1 \\ \tau_2 \text{ if } y'_i \geq \tau_2. \end{array} \tag{5.7}$$

The censoring-rate equations are

$$\Pr(y_i \leq \tau_1 \,|\, x_i) = 1 - \Phi((x_i\beta - \tau_1)/\sigma), \tag{5.8}$$

and

$$\Pr(y_i \geq \tau_2 \,|\, x_i) = \Phi((x_i\beta - \tau_2)/\sigma). \tag{5.9}$$

A Type II model is a sample selection model. A latent variable has a threshold value determining whether the dependent variable is observed or not. That is, an initial decision or filter is applied to determine whether the dependent variable gets past its threshold τ. A mere artifact of measurement (e.g., truncating the recording of income or age) leads to the censored regression model, whereas many economic models of discretionary expenditure involve this kind of initial decision and therefore require sample selection models. The basic equations for the Type II model are

$$y_{oi} = \begin{array}{l} y'_{oi} \text{ if } y'_{si} > \tau \\ \tau \text{ if } y'_{si} \leq \tau, \end{array} \tag{5.10}$$

with y'_{si} denoting the sample selection latent variable and y'_{oi} denoting the outcome latent variable, where

$$y'_{si} = x_{si}\beta_s + e_{si} \tag{5.11}$$

and

$$y'_{oi} = x_{oi}\beta_o + e_{oi}. \tag{5.12}$$

The $\{e_{si}, e_{oi}\}$ are sampled from a bivariate normal distribution with means 0, variances σ_s and σ_o, and covariance σ_{so}. The likelihood function for this model therefore is

$$L_i = \prod_{y_{oi}=\tau} F_s(\tau) \prod_{y_{oi}\neq\tau} f_o(y_{oi}\,|\,y_{si} > \tau)(1 - F_s(\tau)). \tag{5.13}$$

Generally, it is assumed that only the sign of $y'_{si} - \tau$ is observed and that $y'_{oi} = \tau$ simply signifies that $y'_{si} \leq \tau$. That is, there is an indicator variable that tells whether y'_{si} falls below or above a criterion threshold. So, conventionally, $\tau = 0$ and the observed sample often consists of $\{v_i, y_{oi}\}$, where $v_i = 0$ if $y'_{si} \leq \tau$ and $v_i = 1$ if $y'_{si} > \tau$. A comprehensive treatment of sample selection models is beyond our scope, so we will not venture further into that topic here.

Tobit model Types III to V expand on the ideas introduced via Types I and II. In the Type III Tobit, the selection variable is observed only when it passes its threshold. Thus, equation (5.10) becomes

$$
\begin{aligned}
y'_{si} &= \begin{array}{l} y'_{si} \text{ if } y'_{si} > \tau_s \\ \tau_s \text{ if } y'_{si} \leq \tau_s \end{array} \\
y'_{oi} &= \begin{array}{l} y'_{oi} \text{ if } y'_{si} > \tau_s \\ \tau_o \text{ if } y'_{si} \leq \tau_s. \end{array}
\end{aligned}
\tag{5.14}
$$

A typical nonexperimental example is the prediction of the amount on a credit card account during the current month, conditional on any previous debt on the card not entirely paid off before the start of the month. Thus, there are two submodels, one for the amount of remaining debt from the preceding month and another for the amount in the account during the current month. For an experimental example, consider a face-recognition experiment in which the submodels predict response times of participants and the percentage of correct face recognitions, conditional on their rate of correct recognitions exceeding a threshold (e.g., chance-level performance).

The Type IV model adds another outcome variable to the Type III model, but this variable is observed when the selection variable falls below its criterion threshold (i.e., the opposite criterion to the first outcome variable). Consider an experiment in which each participant is presented with a series of choices between two types of puzzles to solve (A and B, say). The submodels predict the percentage of trials on which B-puzzles are chosen. If this percentage exceeds 50%, then the outcome variable observed is the amount of time the participant spends solving B-puzzles; otherwise, the outcome variable observed is the amount of time spent solving A-puzzles. Finally, Type V models differ from Type IV only in that the selection variable is a binary variable as with Type II models. Thus, in our puzzle experiment, participants are asked to indicate an initial preference for A- or B-puzzles and thereafter are given only their preferred kind to solve. The outcome variable is the amount of time spent solving either A-puzzles or B-puzzles, depending on the initial choice.

We now briefly turn to estimation and model diagnostics. As many modelers have pointed out (e.g., Amemiya, 1984), and as we already have observed, not only are OLS estimators biased under censoring, but they are also biased even when restricted to the uncensored observations. Maximum likelihood (ML) estimation of the Gaussian Tobit model, on the other hand, is consistent and also log-concave (the latter implies that standard iterative methods always will converge to the global maximum of the log-likelihood). ML therefore is understandably the default estimation technique for these models.

The tools for Tobit model evaluation and diagnostics consist of those employed for assessing ML-estimated models generally, including the kinds described in Chapter 1. The usual caveats in ML estimation apply here, such as reestimating a model using different starting values. Examining residuals to check whether they are distributed appropriately, identifying any highly influential observations or outliers (e.g., via cross-validation), and checking the correlations between parameter estimates are all key tests of any Tobit model.

5.3 Tobit Model Example

This example uses data from participants recruited by Crowdflower, a crowd-sourcing platform similar in some ways to Mechanical Turk. A Crowdflower sample of 372 adult American participants (mean age = 37.2, SD = 12.8, 193 females) whose data were collected by the authors in mid-2017 included responses to the items on a 12-item "social and economic conservatism scale" (Everett, 2013). This battery has two subscales, one measuring social conservatism and the other measuring economic conservatism. The correlation between the two subscales in this sample is .439. Each item in the battery consists of a single word or phrase, such as "welfare benefits" or "traditional values," and respondents are instructed, "Please indicate the extent to which you feel positive or negative towards each issue. Scores of 0 indicate greater negativity, and scores of 100 indicate greater positivity. Scores of 50 indicate that you feel neutral about the issue." The verbal anchors at the ends of the 0 to 100 scale are *completely negative* and *completely positive*, but these arguably do not constitute absolute bounds. Therefore, scores of 0 or 100 on these items may be regarded as censored scores.

One item in the battery is "gun ownership," a strongly polarizing issue in the United States at the time of this survey. Participant ratings reflected this by exhibiting censoring at both ends of the scale, as can

Figure 5.2   Gun Ownership Histogram

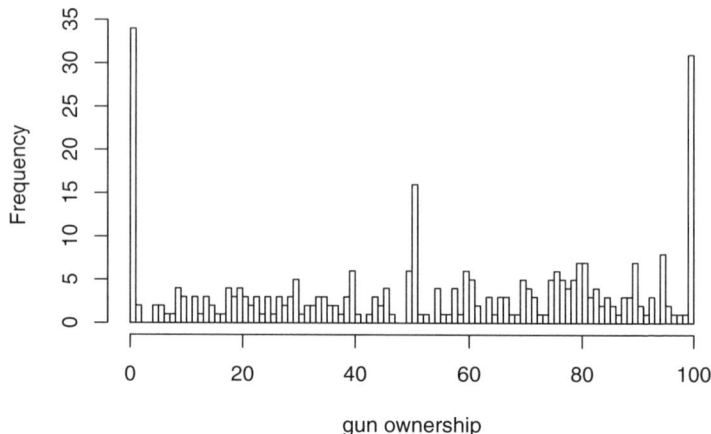

gun ownership

be seen in Figure 5.2. Gun ownership attitude ratings may be modeled via a doubly censored Tobit model (i.e., a Tobit model with censoring at a lower and an upper bound). A more sophisticated approach would be to estimate a mixture model combining the doubly censored distribution with another component distribution to handle the peak at 50. However, for illustrative purposes, we will proceed with just the censored regression model.

We will use political affiliation to predict ratings on the gun ownership item. So, our first model's regression equation is

$$y_i = \beta_0 + \beta_I I_i + \beta_N N_i + \beta_R R_i, \tag{5.15}$$

where I, N, and R are dummy variables taking values 1 or 0 depending on whether a case falls into the Independent, No Preference, or Republican categories, respectively. Thus, β_0 gives the expected value of the Democrats' ratings, and the other coefficients are the expected deviations of the other groups' ratings from the Democrats' mean.

Table 5.1 compares the coefficients in the Tobit model with those obtained by the OLS regression model. The intercept corresponds to the expected value of the Democrats' rating, and it is about 2.4 points lower in the Tobit model. The Tobit coefficients for the Independents and Republicans (these are the significant coefficients), on the other hand, are larger, by 3.7 and 5.7 points, respectively. The overall effect of these differences is to locate the predicted values of the Independents and Republicans farther away from the Democrats in the Tobit

Table 5.1 Maximum Likelihood Model Estimates

Coefficient	OLS	Tobit
β_0	38.510	36.116
β_I	21.526	25.230
β_N	4.468	4.248
β_R	31.607	37.293

model than in the OLS model, which is what we should expect when censoring is taken into account. The OLS model underestimates the extent to which these political groups are in disagreement regarding gun ownership.

There are 30 censored cases at 0 and 29 at 100, so out of the 321 cases in the analysis, the total censoring rate is 18.38%, with the lower-censoring rate 9.03% and the upper-censoring rate 9.35%. The lower- and upper-censoring rates for each of the political groups can be computed from Table 5.1 and equations (5.8) and (5.9). For example, the Democrats' lower-censoring rate is $1 - \Phi((x_i\beta - 0)/\sigma_i) = 1 - \Phi((36.116)/35.899) = .1572$ (i.e., 15.72%). Their upper-censoring rate is $\Phi((x_i\beta - 100)/\sigma_i) = \Phi((36.116 - 100)/35.899) = .0376$, or 3.76%. By contrast, the Republican lower-censoring rate is $1 - \Phi((36.116 + 37.293)/35.899) = .0204$, or 2.04%, and their upper-censoring rate is $\Phi((36.116 + 37.293 - 100)/35.899) = .2294$, or 22.94%.

The Tobit model-censoring predictions are reasonably accurate for the Democrats (empirical lower- and upper-censoring rates are 14.29% and 2.04%), Republicans (empirical lower and upper rates are 2.94% and 17.65%), and Independents (predicted rates are 4.37% and 14.08%; empirical rates are 3.67% and 10.09%). However, it underestimates how polarized the No Preference group is (predicted rates are 13.04% and 4.83%, but empirical rates are 21.73% and 13.04%).

Relevant diagnostics for this type of model include examining the parameter estimates correlation matrix and the model residuals. Inspection of the correlation matrix reveals that the strongest correlations are between the intercept and the other coefficients, with the strongest of these being –.971, large enough to be concerning. Figure 5.3 displays a normal Q-Q plot, whose tails indicate that the residuals are slightly kurtosed. This, combined with our observations regarding the model's poor reproduction of censoring rates in the No Preference group, suggests that there may be nonnormality and/or some heteroscedasticity

Figure 5.3 Political Group Model Tobit Residuals

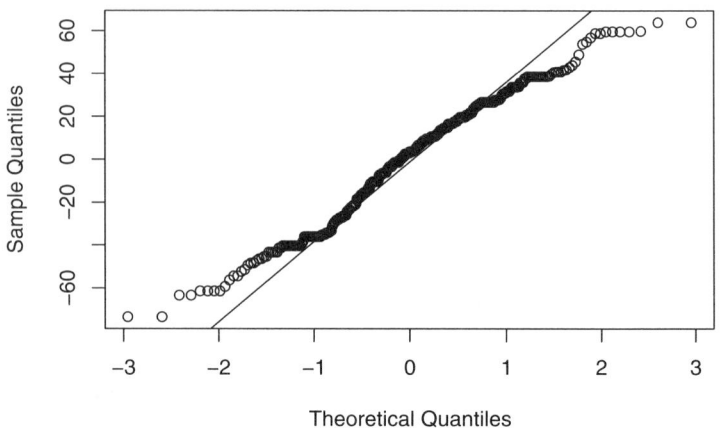

in the data. To conclude this chapter, we extend our example by fitting heteroscedastic non-Gaussian models to the data.

5.4 Heteroscedastic and Non-Gaussian Tobit Models

What are motivations for considering heteroscedastic and/or non-Gaussian Tobit models? Heteroscedasticity presents a problem for Tobit models, for two reasons. The first reason is one that applies to any misspecified model, namely, that we are ignoring what may be important effects of covariates on variance. A second reason is that under heteroscedasticity, Tobit estimates are inconsistent.

There are several ways to test for heteroscedasticity, most of which involve different varieties of a score test. Holden (2011) provides a review and comparisons among these via Monte Carlo studies, concluding that the best-performing estimator is the statistic first presented by Jarque (1981). There are two effective responses to heteroscedasticity. One approach is to use a censored least absolute deviations estimator (Powell, 1984), whose estimators have been shown to be consistent under heteroscedasticity (also nonnormality) of unknown form. The other is to model the heteroscedasticity via an appropriate link function in a dispersion submodel. The choice between these alternatives depends in good part on the researcher's goals (e.g., whether modeling the heteroscedasticity is of theoretical or substantive interest).

The heteroscedastic Tobit model requires that σ in equation (5.3) for the error term e_i in equation (5.2) be permitted to vary, and a general model for it is

$$h(\sigma_i) = z_i \delta, \qquad (5.16)$$

where h is a link function that guarantees positive predicted values, z_i is a vector of predictors, and δ is a vector of coefficients. A reasonable choice for h often is the log. Likewise, a practical approach to testing for heteroscedasticity is simply to estimate an appropriate model and compare it with its homoscedastic counterpart.

Departures from normality also are problematic for the Tobit model, again both because a Gaussian model is then a misspecified model and because they render the maximum likelihood estimator inconsistent. Holden (2004) compares several tests for normality, showing that one popular test, presented by Chesher and Irish (1987) and recommended in some texts such as Breen (1996), has excessive Type I error rates. Holden's own recommendation is the Bera-Jarque-Lee test (Bera, Jarque, & Lee, 1984). As with heteroscedasticity, a pragmatic approach to dealing with nonnormality involves choosing between estimating a semiparametric model such as a Powell-type estimator or a parametric model specifying a distribution. We will take a parametric model approach here for illustrative purposes.

After analyzing the gun ownership attitude data with a Tobit model, we surmised that there could be heteroscedasticity and/or nonnormality in the data. We now put this to the test by comparing a heteroscedastic non-Gaussian model with the homoscedastic model. As an example of a non-Gaussian Tobit model, we extend the gun ownership example, combining the economic conservatism scale (Everett, 2013) and political affiliation as predictors. The relatively fat tails in the residuals of the Tobit model in Section 5.4 suggest that a distribution with heavier tails than the Gaussian might be more effective for the gun ownership attitude data. The logistic distribution is similar to a t distribution with four degrees of freedom and is available as a Tobit model in some statistical packages. Therefore, we will compare heteroscedastic logistic and homoscedastic Gaussian models and use them to address the question of whether economic conservatism mediates the relationship between political affiliation and gun ownership attitudes.

Both the Gaussian and logistic Tobit models yield significant positive coefficients for the Economic Conservatism Scale (ECS) in predicting the gun ownership attitude scale and nonsignificant coefficients for the political affiliation groups. A likelihood ratio test comparing each of these models with their counterparts when political affiliation is removed yields $\chi_3^2 = 4.439$ ($p = .218$) for the Gaussian homoscedastic and $\chi_6^2 = 7.696$ ($p = .261$) for the logistic heteroscedastic model. The ECS location submodel coefficients in the model with only ECS as a predictor are 1.791 ($z = 16.874, p < .0001$) for the Gaussian and 1.820 ($z = 16.717, p < .0001$) for the logistic.

The two models agree that economic conservatism mediates the relationship between political affiliation and gun ownership attitudes, and their evaluations of the effect of economic conservatism are similar. The heteroscedastic Tobit for these data includes a term in the dispersion submodel for the ECS scale, and while its coefficient has a small magnitude (–0.008 for the Gaussian and logistic models), the likelihood ratio test comparing this model with a homoscedastic logistic Tobit shows a significant difference in fit ($\chi_1^2 = 5.051, p = .025$). The negative coefficient tells us that for higher values of ECS, there is less dispersion in the gun ownership ratings.

The AIC values for these models confirm that the heteroscedastic models are not likely to be overfitting. The AIC for the homoscedastic logistic model is 2,569.567, whereas for the heteroscedastic model, it is 2,566.517. Likewise, for the Gaussian homoscedastic model, the AIC is 2,583.537, whereas for its heteroscedastic counterpart, the AIC is 2,579.383.

Do we require a non-Gaussian heteroscedastic model here, or would a Gaussian heteroscedastic model suffice? A heteroscedastic t distribution Tobit model with degrees of freedom as a free parameter yields an estimated $df = 3.796$, so this distribution is very similar to the logistic distribution and certainly heavier tailed than the Gaussian distribution. However, these models all have nearly identical correlations between their fitted values and those of the dependent variable (about 0.707). Thus, some evidence favors the non-Gaussian models on grounds of goodness of fit, although for predictive purposes, a Gaussian model would do as well. The important augmentation is incorporating heteroscedasticity into our Tobit model.

Model diagnostics also indicate that the heteroscedastic logistic model behaves reasonably well. There is one worrying correlation in the parameter estimates correlation matrix, namely, a –.977 correlation between the intercept and ECS coefficient in the location submodel. However, all other correlations' magnitudes are below .33. Finally, as

Figure 5.4 Heteroscedastic Logistic Model Residuals

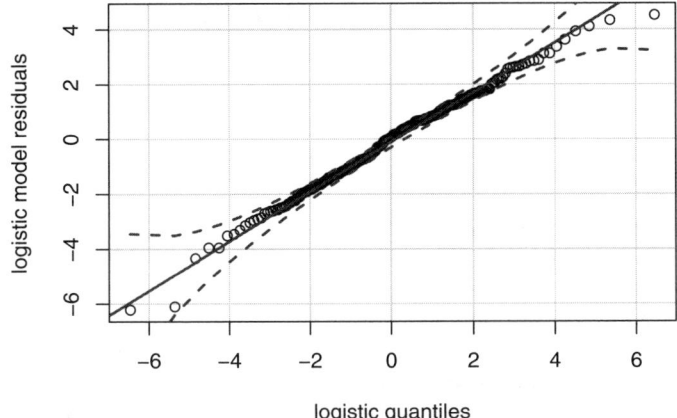

the Q-Q plot in Figure 5.4 shows, the residuals fit their theoretical distribution well. We may conclude that the heteroscedastic logistic Tobit model is suitable and perhaps slightly superior to the heteroscedastic Gaussian Tobit model in this application. Nevertheless, we hasten to emphasize the point that there often is no indubitably superior model in situations such as this one. Instead, a typical outcome is a set of serviceable models, with the researcher's final choice often being directed by pragmatic considerations.

CHAPTER 6. EXTENSIONS AND CONCLUSIONS

6.1 Extensions and a General Framework

In this book, we intended to provide researchers and students with an introduction to methods for analyzing continuous variables with bounds. We have focused on three types of bounds: absolute scale limits, censored scores, and sample truncation. We have limited our introductory treatment to variables that are "pure" examples of each of these types of bounds, data that consist of independent and identically distributed observations, and parameter estimation and model diagnostic techniques oriented around maximum likelihood estimation. Real research projects often present researchers with more complex kinds of data, so in this chapter, we provide some glimpses of that broader territory. The next three sections discuss variables in which absolute bounds coexist with censoring or truncation, extensions of several types of models to dealing with multilevel and multivariate data, and modeling bounded variables in the Bayesian framework.

First, however, we integrate models of bounded variables into an overall framework, with the goal of enabling researchers to make informed choices about the kinds of models appropriate for analyzing their data. These choices may incur debates and even some controversy. Briefly, the primary decision the analyst must make concerns the nature of the boundary or boundaries of a variable's distribution and then a secondary decision about the appropriate models for dealing with observations on and off the boundaries. The outcome of the first decision specifies a latent variable but only partly determines the outcome of the second decision about how to model it.

Each type of bound brings with it a distinct latent variable underlying the data. We have already seen that the absolute boundary latent variable model restricts the support for the distribution within the bounds, whereas the censored and truncated latent variable models posit a support for the distribution that extends beyond the bounds. It is worth bearing in mind that there can be "hybrids" (e.g., censoring or truncation of a variable that has one or more absolute bounds). Indeed, the example illustrated in Figure 5.1 in Chapter 5 is a case in point. The distribution is of bank losses, where the loss amounts are recorded only if they exceed $1,000. The boundary at $0 may be regarded as absolute, and the censoring bound occurs at $1,000. We will examine another hybrid of this kind in the next section.

As mentioned in Chapter 1, the nature of a scale boundary sometimes can be contestable, or it may differ depending on the researcher's purposes. Scales constructed by linear combinations of items provide exemplars of both possibilities. In Chapter 1, we presented a relatively straightforward instance, namely, an examination. As we pointed out, the nature of the bounds on examination scores depends on whether we regard the exam questions as exhausting the subject matter or not. We now extend the scope of discussion. Two types of consideration may bear on how a boundary is to be treated: whether the definition of the construct or its operationalization entails a bound with a specified nature and whether a scale boundary also constitutes a population boundary rather than a sample-specific boundary.

Definitional considerations may be relevant when boundary conditions form part of the definition or a construct or its measurement. For instance, a definition of an organization's "income" restricted to money flowing into the organization's accounts imposes a lower bound of 0, whereas if income refers to the net flow of money into and out of those accounts, then negative incomes are admissible observations. However, not all bounds that exist by definition are pertinent to the analyst. Hypothetical bounds that are unattainable may justifiably be ignored. Consider measuring the number of hours in one 30-day month that a person spends shopping online. The lower bound of 0 clearly is relevant, but the upper bound of 720 hours may not be, if it is humanly impossible to shop online continuously for 720 hours.

Now we turn to the question of whether a boundary constitutes a population boundary or a sample-specific boundary. The most common type of scale for which this question is relevant is a scale consisting of a linear combination of scores from a collection of items. A reasonable indication of whether a particular scale's bounds should be treated as absolute or censored is whether its collection of items may be regarded as exhausting the population of such items. For some psychological or physiological disorders, the relevant symptoms form a small enough collection that they may all be included in a scale measuring the disorder. Some kinds of anxiety disorders arguably are like this, for instance. Other disorders may not be exhaustible in this way, either because their symptom list is too large or because it is not possible to obtain an exhaustive list. Delusional thinking is an example of the latter problem, both because we can never know when a new delusion may present itself and also due to the difficulties inherent in defining the concept of a delusion. Many culturally and/or historically specific attitude scales' bounds also should be regarded as censoring rather than absolute for similar reasons.

The second decision for the analyst is the type of statistical model to apply to the latent variable model, given the decision about the nature of the bounds. A latent variable with absolute bounds may be modeled by a single distribution whose support is defined by the bounds, as in regression models for single-bounded variables with "life expectancy" distributions or models for doubly bounded variables with beta or cdf-quantile distributions. However, such a latent variable also may be analyzed with sample selection or boundary inflation models. The analyst's choice may be based on any of three considerations. First, there may be cases at the boundaries but the appropriate model is undefined at the boundaries (e.g., beta regression models are not generally able to deal with 0s or 1s). Second, even if a model is defined at the bounds of its support, the analyst may prefer to treat boundary cases as qualitatively distinct from cases off the boundaries, in which case a hurdle model or a sample selection model may be suitable. Third, even if boundary cases are not regarded as qualitatively different, the analyst may posit that the processes generating the data yield an excess of boundary cases beyond what is expected according to the analyst's distributional model, so that a boundary inflation model is appropriate.

Likewise, we have seen that the analyst has a choice between at least two kinds of models for dealing with censoring, as in Amemiya's (1984) typology of Tobit models. For instance, recall that the Type I model assumes fixed lower and/or upper threshold(s) beyond which an observation's true score is known only to have exceeded or fallen below a threshold. The Type II, or sample selection, model posits an additional latent variable with a threshold determining whether an observation's score is known or not.

A fairly commonplace type of data that we have not dealt with is interval-valued data, as in classification of observations into contiguous bins such as age groups or income brackets. These data are "censored" in the sense that we do not know someone's exact age if he or she falls in the 20- to 30-year age group. The interval boundaries may be fixed, as in the examples thus far, or random variables in their own right as in the random threshold Tobit model briefly described in Chapter 5. A typical example of varying-boundary interval data is the elicitation of lower and upper estimates of some quantity from a sample of judges. For instance, a 27-nation study of lay numerical translations of verbal probability expressions used in the Fourth Report by the Intergovernmental Panel on Climate Change (Budescu, Por, Broomell, & Smithson, 2014) elicited lower, upper, and "best" estimates from participants.

6.2   Absolute Bounds and Censoring

Variables that have both absolute bounds and censoring typically arise as a by-product of a format for recording data that has limited precision and therefore uses intervals instead of precise values. Quite often, this issue is ignored by researchers, usually for pragmatic reasons. For instance, a common practice is to treat interval data as if they are precise by using the midpoints of the intervals as if they are the actual data. In this section, we will consider an example of a doubly bounded variable that also is doubly censored.

The Crowdflower survey conducted by the authors in 2017 (described in Section 5.2) included an experimental component, whereby participants were randomly assigned to receive one of these two questions: "In the next 5,000 years, what is the probability that the human species will become extinct?" or "Within how many years' time do you expect that the human species will become extinct?" Of the sample of 372 adult Americans, 330 cases were determined to be usable for the extinction probability estimates.

The response formats asked respondents to choose a range (e.g., "between 1 in 100 and 1 in 500 chances" vs. "100 to 500 years from now") and then a specific value from a list within that range (e.g., 1 in 300 chances vs. 300 years). These alternatives are equated with one another by taking expectations of the probability to arrive at the expected number of years before an extinction event. For instance, an event whose probability is 1 in 300 chances within a given year would be expected to occur once every 300 years.

Both response formats had boundaries on the probabilities that were above 0 and below 1. The lowest probability that could be assigned was less than 1 in 55,000 chances versus more than 55,000 years (.000018), and the highest was more than 1 in 10 chances versus less than 10 years (.1). Thus, these are censored scores. For example, a participant recording his or her probability estimate as "more than 1 in 10 chances" has nominated a probability somewhere between .1 and 1. Of the 330 responses, the censoring rates were 44 (13.33%) at the lower boundary and 10 (3.03%) at the upper boundary. Neither of these rates differed significantly between the two response formats ($\chi^2_1 = 2.95$, $p = .086$

and $\chi_1^2 = 2.22$, $p = .136$, respectively), so we will ignore the format from here on and model these data as a random sample from a doubly bounded random variable with censoring inside its support.

Suppose we want to compare the results obtained if we model means and variance with the results from modeling medians and other quantiles. Our motivation for this might be that we are concerned about the skew in our data and therefore wish to see how "robust" predictor effects are if we use alternative measures of central tendency such as the mean versus the median. For illustrative purposes, the candidate distributions for our models will be a censored beta and censored logit-logistic distribution (recall that the logit-logistic is a member of the CDF-quantile family).

Our goal is to ascertain whether political affiliation is related to judged probabilities of human extinction. Political affiliation was measured in two ways. First, participants were asked to assign themselves to the Democrat, Republican, Independent, or No-Preference affiliation. Second, they completed Everett's (2013) social and economic conservatism scales. Initial models included self-assigned political affiliation, social conservatism score, and economic conservatism score in both the location and dispersion submodels. Backward elimination was then employed to arrive at the final models.

We begin with beta regression. The censored beta regression model includes political affiliation and social conservatism in the location submodel and political affiliation in the precision submodel. The political affiliation part of the model used Republicans as the base group against which the other affiliations were compared. Table 6.1 shows the relevant regression coefficients, standard errors, and confidence intervals. In the location submodel, the political affiliation effect was due to the Independents probability ratings being lower than Republicans, whereas higher scores on the social conservatism scale predicted lower probability ratings. In the precision submodel, the Democrats and Independents probabilities exhibited less dispersion than those of the Republicans.

Now we turn to CDF-quantile regression. The censored logit-logistic regression model coefficients, standard errors, and confidence intervals are shown in Table 6.2. This model turns out to be simpler than its beta regression counterpart, with the social conservatism scale accounting for both location and dispersion effects. Missing are effects from political affiliation. As in the beta regression, higher social conservatism scores predict lower judged probabilities of human extinction, but unlike the beta regression model, higher social conservatism scores also predict greater dispersion among these probability estimates.

Table 6.1 Censored Beta Regression Model Estimates

			Location submodel			
	Coefficient	Estimate	*SE*	z	2.5%	97.5%
Intercept	β_0	−3.630	0.378		−4.372	−2.888
Democrat	β_1	−0.638	0.348	−1.822	−1.324	0.048
Independent	β_2	−0.973	0.339	−2.976	−1.655	−0.291
No Preference	β_3	−0.276	0.439	−0.628	−1.137	0.585
Social conservatism	β_4	−0.010	0.003	−2.900	−0.016	−0.003
			Precision submodel			
	Coefficient	Estimate	*SE*	z	2.5%	97.5%
Intercept	δ_0	2.777	0.328		2.135	3.420
Democrat	δ_1	0.930	0.402	2.315	0.143	1.717
Independent	δ_2	1.146	0.402	2.853	0.359	1.934
No Preference	δ_3	0.326	0.508	0.642	−0.669	1.322

Table 6.2 Censored Logit-Logistic Model Estimates

			Location submodel			
	Coefficient	Estimate	*SE*	z	2.5%	97.5%
Intercept	β_0	−5.386	0.487		−6.341	−4.431
Social conservatism	β_1	−0.0250	0.008	−3.255	−0.040	−0.010
			Dispersion submodel			
	Coefficient	Estimate	*SE*	z	2.5%	97.5%
Intercept	δ_0	0.076	0.159		−0.235	0.388
Social conservatism	δ_1	0.005	0.002	2.098	0.000	0.010

It is worthwhile to inquire into why these two models differ on the predictors they have retained, even bearing in mind that each is estimating a different set of parameters. The beta model is estimating the conditional mean and precision, while the logit-logistic model is estimating conditional quantiles. We can gain an intuitive understanding

Figure 6.1   Political Affiliation by Probability of Extinction Estimates

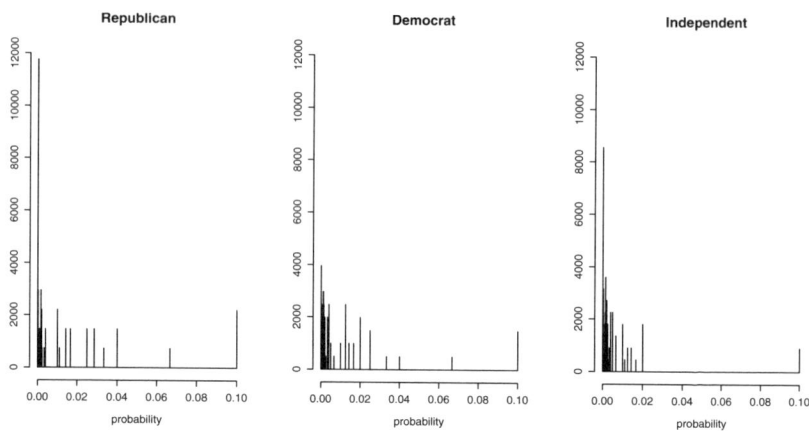

of these models by examining the distributions of probability estimates for three of the political groups, as displayed in Figure 6.1. These graphs have been drawn so that the densities accumulate to 1, thereby removing the effects of the different group sizes.

It is clear that the Independents tend to have lower probability estimates than the Democrats or Republicans and that the Republicans are more strongly polarized (greater censoring rates) than the Democrats and Independents. The most likely explanation for the differences between the beta and CDF-quantile (logit-logistic) models resides in the combination of censoring and the outliers in the right tail. These influence the beta model more than the CDF-quantile model because the location submodel in the former is predicting the conditional mean, whereas the location submodel in the latter is predicting the conditional median.

Circumstantial evidence for this explanation resides in the expected values that our models provide. Starting with the censored beta regression, from Table 6.1, the mean of the social conservatism scale is 64.337, so the predicted mean (taking censoring into account) of the Republicans is

$\hat{\pi}_R = \exp(-3.630-0.010*64.337)/(1 + \exp(-3.630 - 0.010*64.337))$
$= 0.014$. The predicted mean for the Democrats is
$\hat{\pi}_D = \exp(-3.630 - 0.010*64.337)/(1+\exp(-3.630 - 0.638 - 0.010* 64.337)) = 0.007$. Likewise, inserting their respective coefficients into

this function reveals that the predicted means for the Independent and No Preference groups are $\hat{\pi}_I = 0.005$ and $\hat{\pi}_N = 0.010$, respectively. If we ignore censoring and run a conventional beta regression, then the predicted means are $\hat{\pi}_R = 0.011$, $\hat{\pi}_D = 0.007$, $\hat{\pi}_I = 0.005$, and $\hat{\pi}_N = 0.009$. So taking censoring into account pulls the Republican and No Preference means higher and therefore further away from the means for the Democrats and Independents. This is because the upper-censored observations are being more strongly underestimated than the lower-censored observations are overestimated.

Turning now to the censored logit-logistic model, at the mean of the social conservatism scale, the predicted median is $\hat{y}^{(0.5)} = 1/(1 + \exp(5.323 + 0.027 * 64.337)) = 0.00086$. If we ignore censoring and run a conventional logit-logistic model, the predicted median turns out to be $\hat{y}^{(0.5)} = 1/(1 + \exp(5.395 + 0.025 * 64.337)) = 0.00091$, so taking censoring into account hardly affects the median. This is because the median is generally little influenced by extreme scores.

However, before leaping to the conclusion that censoring does not matter for modeling quantiles, we should take a look at the impact of censoring on quantiles other than the median. For the logit-logistic model, recall that in Chapter 4, we displayed the quantile function in equation 4.13:

$$G^{-1}(\gamma, \mu, \sigma) = \frac{1}{1 + \exp\left(-\left(\frac{\mu}{\sigma} - \log(\frac{1-\gamma}{\gamma})\right)\sigma\right)}.$$

We will use this equation to see what impact censoring has on the first and third quartiles. Substituting the β_j and δ_j coefficients from Table 6.2 into this equation, the predicted first quartile is $\hat{y}^{(0.25)} = 0.00015$ and the predicted third quartile is $\hat{y}^{(0.75)} = 0.00498$. Now, we already have seen that the logit-logistic model ignoring censoring has an intercept of –5.394 and a coefficient for social conservatism of –0.025. These differ little from the intercept and coefficient in the censored logit-logistic model, so the predicted median also differs little from its censored logit-logistic model counterpart. However, the dispersion submodel ignoring censoring has an intercept of 0.055 and a coefficient for social conservatism of 0.004. These differ fairly substantially from their counterparts in Table 6.2. For this model, the predicted first quartile is $\hat{y}^{(0.25)} = 0.00020$ and the predicted third quartile is $\hat{y}^{(0.75)} = 0.00406$. Ignoring censoring brings the predicted first and third quartiles closer together. The difference between these predicted quartiles (i.e., the interquartile range) is 0.00484 if we take censoring into account but only 0.00386 if we ignore censoring. This example illustrates that censoring often has stronger effects on dispersion than on central tendency and therefore has its largest impact on extreme quantiles (i.e., those that are far from the median).

6.3 Multilevel and Multivariate Models

Bounded variables present important challenges and considerations for constructing multilevel and multivariate models. Some of these considerations have not been well studied, so they pose genuine unknowns for modelers. In this section, we briefly review the alternatives and issues regarding such models pertaining to dependent variables with boundaries. We also present a detailed example of a random intercept model in which the dependent variable is boundary inflated.

In the absence of cases on the boundaries themselves, models of multivariate dependent variables with bounds do not require any additional concepts or technical considerations to models of collections of dependent variables generally. A popular approach to modeling such collections is to use copulas, and this approach is well suited to many bounded-variable distributions. Briefly, a copula is a multivariate probability distribution whose marginals have uniform distributions. A good introduction to this topic is Nelsen's (1999) book. Sklar's (1959) theorem states that any multivariate distribution can be translated into a copula describing the dependency structure among its variables and the variables' marginal distribution functions. Let $F(x_1, \cdots, x_J) = \Pr(X_1 \leq x_1, \cdots, X_J \leq x_J)$ be a multivariate CDF. Then there is a copula, C, such that $F(x_1, \cdots, x_J) = C(F_1(x_1), \cdots, F_J(x_J))$, where $F_j(x_j)$ is the jth marginal CDF. Moreover, if the domains of the CDFs are bounded and the CDFs are continuous, then the copula is unique. The converse holds as well: Given a J-dimensional copula C and a collection, $F_1, \cdots, F_J$, of marginal CDFs, $C(F_1(x_1), \cdots, F_J(x_J))$ defines a multivariate CDF.

The copula is understandably a popular way of modeling multivariate distributions, because the marginal distributions and dependency structure can be modeled separately. Distributions with invertible CDFs, such as the CDF-quantile family, provide completely tractable multivariate models via copulas. Moreover, some software environments have the resources to construct and test these models. Shou and Smithson (2019) include an example of such a copula model as a demonstration of how their cdfquantreg package can work with other packages in the R environment.

Multivariate dependent-variable structures pose interesting problems when boundary cases are present, either due to censoring or in collections of boundary-inflated variables where hurdle or boundary inflation models are adopted to deal with boundary cases. For instance, censoring in one variable may depend on censoring in another. The question of how to model the dependency structures of such variables is an active research topic. There is a sizable literature on multivariate Tobit models, and one of the more popular approaches is to use copulas to model the dependence structure (Huang, Sloan, & Adamache, 1987). Early versions of the copula approach restricted errors to being normally distributed (Trivedi & Zimmer, 2007), but more recent models modify or relax this assumption (e.g., Gupta & Gupta, 2008; Louzada & Ferreira, 2015).

Likewise, a literature has sprung up around the problem of multivariate boundary-inflated distributions, with the earliest treatment apparently due to Aitchison (1955). Bascoul-Mollevi, Gourgo-Bourgade, & Kramar (2005) developed two-part statistics for analyzing paired data that are zero inflated. The statistics included a test of proportions for the counts of the zeros and a statistic comparing the means. Daoud (2007) extended these two-part statistics to comparisons among k samples with zero-inflated distributions. Pimentel, Niewiadomska-Bugaj, & Wang (2015) proposed modified estimators of Kendall's τ and Spearman's ρ, exploiting the fact that bivariate boundary inflation is a special case of tied scores. Much of the literature on both multivariate censoring and boundary inflation is devoted to computational issues regarding parameter estimation.

Researchers estimating multilevel models face decisions regarding where and how to insert random-effects terms. Depending on the structure of the data, it may make sense to have random-effects terms in the censoring component or hurdle component of a model. Random-effects censoring, hurdles, and boundary inflation are relatively straightforward conceptually, but they may be difficult to estimate without a large number of observations per group. Rabe-Hesketh, Yang, & Pickles (2001) provide a general approach to multilevel models involving censoring, and Wong, Zeng, & Lin (2017) discuss estimation procedures for structural equations models with censored data. Dependent variables with absolute boundaries present some difficulties regarding estimation of random-effects terms in dispersion submodels. Verkuilen and Smithson (2012) describe such problems for mixed beta regression models, observing that they tend to occur when the bulk of the data is close to either the 0 or 1 boundary or when the precisions are low.

We now turn to an example illustrating a multilevel model that includes doubly bounded variables with ceiling effects and employs "1-inflated" beta distribution models to account for them. The data are from the World Values Survey data set employed for examples earlier in this book. The questions are from a group of items that ask participants to rate the degrees to which various characteristics of a polity are an essential component of a democracy. The rating scale is 1 to 10, where 1 denotes *not at all an essential characteristic of democracy* and 10 denotes *an essential characteristic of democracy*. Six items from the battery of nine selected for our example are the following:

- X132. Religious authorities ultimately interpret the laws.

- X133. People choose their leaders in free elections.

- X135. The army takes over when government is incompetent.

- X136. Civil rights protect people from state oppression.

- X137. The state makes people's incomes equal.

- X139. Women have the same rights as men.

X133, X136, and X139 are strongly right-censored because most survey respondents agree that these are essential to democracies. X132, X135, and X137 are strongly left-censored because most respondents agree that these are not essential to democracies. For modeling purposes, we have reverse-scored the latter three items and divided the scores by 10 so that all six items are now 1-inflated variables whose scores are in the (0,1] interval. Figure 6.2 displays the frequency distributions of the six items, demonstrating that all of them are strongly 1-inflated.

We have selected one predictor for inclusion in our model, namely, political orientation, as elicited by this question: "In political matters, people talk of 'the left' and 'the right.' How would you place your views on this scale, generally speaking?" The 1 to 10 scale is anchored with *left* at 1 and *right* at 10. We will test two hypotheses: first, that a stronger left-wing orientation predicts a higher probability of returning a score at the boundary (i.e., a score of 1) on all items and, second, that a stronger left-wing orientation predicts a higher score for the scores below 1.

Before fitting such a model, we must consider how to deal with random-effects terms. There are potentially three submodels in which

Figure 6.2   Ratings of Six Candidate Characteristics

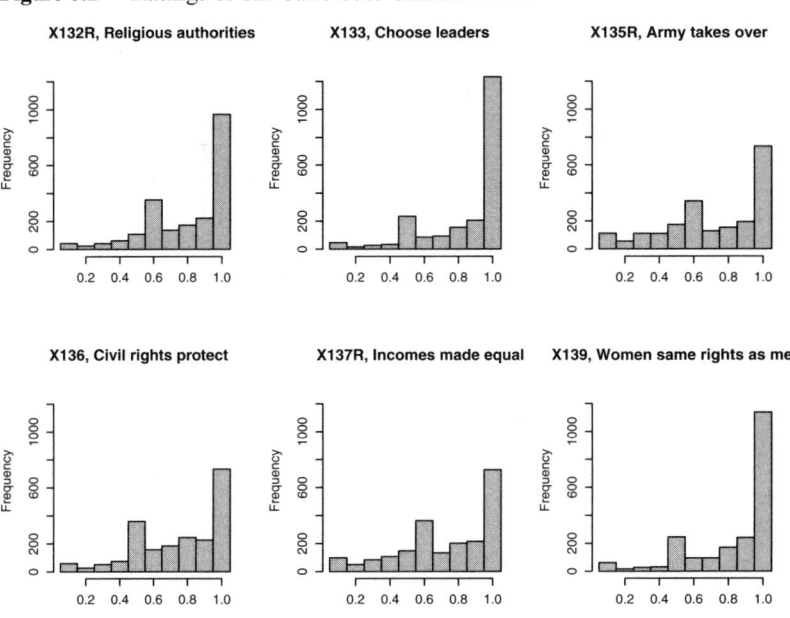

random-effects terms could be included: the location, precision, and hurdle submodels. Each case has at most six observations, so there are not enough observations per case for a stable model that estimates three random intercepts (typically we would need at least several observations to dedicate to each random term). To begin, we restrict the model to a random intercept in the location submodel. This restriction results in models that converge successfully.

The best model identifies item main effects in all three submodels, a main effect of political orientation in the precision submodel, and an effect of political orientation that is moderated by item effects in the location submodel. Starting with the hurdle submodel, inspection of Figure 6.2 suggests that X133 and X139 are the most strongly 1-inflated items, followed by X132 (reverse-scored), and then the remaining three. This is exactly what the hurdle submodel indicates, and it reproduces the 1-inflation rates perfectly.

Table 6.3 displays the coefficients for the hurdle model. The empirical probabilities of 1s for the six items are (0.455, 0.580, 0.348, 0.346, 0.342, 0.537). We can readily verify that the hurdle model reproduces these. The v_0 coefficient yields $\exp(-0.181)/(1 + \exp(-0.181)) = 0.455$ (the 1's

Table 6.3   Hurdle Model Coefficients

	Coefficient	Estimate	SE	z	2.5%	97.5%
Intercept	v_0	−0.181	0.044	–	–	–
X133	v_1	0.503	0.062	8.129	0.386	0.620
X135R	v_2	−0.447	0.063	−7.077	−0.550	−0.343
X136	v_3	−0.454	0.063	7.192	−0.559	−0.349
X137R	v_4	−0.475	0.063	−7.514	−0.579	−0.371
X139	v_5	0.328	0.062	5.320	0.214	0.441

Table 6.4   Dispersion Submodel Coefficients

	Coefficient	Estimate	SE	z	2.5%	97.5%
Intercept	η_0	2.602	0.064	–	–	–
X133	η_1	−0.243	0.060	−4.088	−0.360	−0.127
X135R	η_2	−0.461	0.053	−8.688	−0.564	−0.356
X136	η_3	0.035	0.054	0.651	−0.070	0.140
X137R	η_4	−0.413	0.053	−7.769	−0.517	−0.308
X139	η_5	−0.283	0.058	−4.892	−0.396	−0.170
Conservative	η_6	−0.057	0.008	−6.959	−0.074	−0.041

probability for X132R), the v_0 and v_1 coefficients produce $\exp(-0.181 + 0.503)/(1 + \exp(-0.181 + 0.503)) = 0.580$ (the 1's probability for X133), the v_0 and v_2 coefficients produce $\exp(-0.181 - 0.447)/(1 + \exp(-0.181 - 0.447)) = 0.348$ (the 1's probability for X135R), and so on.

The coefficients for the dispersion submodel are displayed in Table 6.4. These are parameters that are used in the gamlss package in R (Stasinopoulos & Rigby, 2007): $\eta = 1/(1 + \phi)$, where ϕ is the precision parameter for the beta distribution introduced in Chapter 3. Larger η therefore indicates greater variance, but we must also bear in mind that the variance of the beta distribution depends on $\mu(1 - \mu)$, and the variance of the 1-inflated beta also depends on the proportion of 1s in the data. The sample variance is 0.065 and the model's estimate is somewhat lower at 0.054. The sample variances for the six items are (0.054, 0.053, 0.079, 0.062, 0.073, 0.057), and the model also produces underestimates of these variances, (0.046, 0.044, 0.069, 0.052, 0.063, 0.049), although these estimates follow the same pattern of relative magnitudes as their sample counterparts. Thus, X135R and X137R have the largest variances even though they do not have the largest η values.

Table 6.5 Location Submodel Coefficients

	Coefficient	Estimate	*SE*	z	2.5%	97.5%
Intercept	β_0	1.148	0.065			
X133	β_1	−0.309	0.105	−2.944	−0.514	−0.103
X135R	β_2	−0.364	0.094	−3.871	−0.548	−0.180
X136	β_3	−0.370	0.086	−4.327	−0.538	−0.202
X137R	β_4	−0.647	0.089	−7.271	−0.821	−0.472
X139	β_5	−0.297	0.102	−2.897	−0.497	−0.096
Conservative	β_6	−0.087	0.011	−8.178	−0.108	−0.066
Conservative * X133	β_7	0.056	0.018	−3.171	0.021	0.090
Conservative * X135R	β_8	−0.003	0.015	−0.171	−0.033	0.028
Conservative * X136	β_9	0.034	0.014	2.443	0.007	0.062
Conservative * X137R	β_{10}	0.063	0.015	4.194	0.035	0.093
Conservative * X139	β_{11}	0.044	0.017	2.633	0.011	0.077

The coefficients for the location submodel are displayed in Table 6.5. The sample means for the six items are (0.804, 0.848, 0.716, 0.752, 0.731, 0.833), and the model estimated means (0.807, 0.853, 0.715, 0.745, 0.728, 0.832) match these fairly closely. According to this submodel, X133 has the highest mean and X135R the lowest, with a more conservative political orientation predicting a lower mean for all items.

Finally, the model also estimates individual participants' means via the random intercept component. The correlation between those estimates and individuals' mean ratings is .537, and Figure 6.3 shows the resulting scatterplot of the empirical means and model predictions. Note that the fan-shaped scatterplot arises from the natural heteroscedasticity of a regression model for a doubly bounded random variable.

6.4 Bayesian Estimation and Modeling

Bayesian statistical methods are increasingly employed throughout the human sciences and are likely to become the standard in some domains. This is a rapidly expanding topic, and at this time, accessible introductions include Kruschke (2014) and McElreath (2015). We present

Figure 6.3   Empirical versus Predicted Means

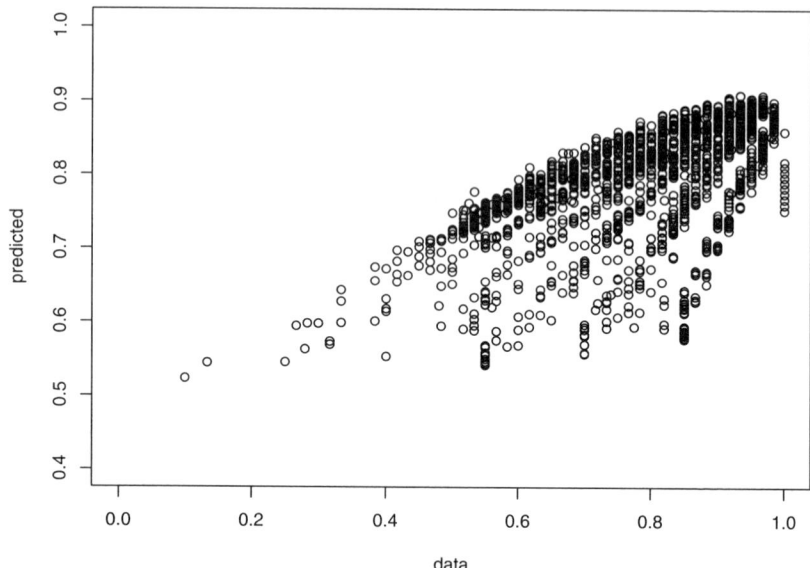

an example using Bayesian Markov chain Monte Carlo (MCMC) to estimate a model of the data from the grammaticality judgment study described in Chapter 3. These data comprise a sample of 247 sentences that were translated into one of the four target foreign languages (Norwegian, Spanish, Chinese, or Japanese), and then back-translated into English via machine translation. For each sentence, 8 to 14 participants rated the "naturalness" of the sentence translations using a slider scale from 1 (*extremely unnatural*) to 100 (*extremely natural*).

In Chapter 3, we modeled the effects of language and sentence length on the mean rating for each sentence using a beta GLM, with language and sentence-length effects in both the location and precision submodels. The supplementary materials for this chapter include a Bayesian MCMC estimation of the same model, showing that it yields similar coefficient values but somewhat larger standard errors. These results are as would be expected from a Bayesian model when compared with its frequentist counterpart.

However, in this section, we use Bayesian MCMC to estimate a variance partition model that would be somewhat difficult to estimate via frequentist methods. Here, instead of treating the mean rating as a precise score as we did in Chapter 3, we include the standard errors of the means for each of the sentences and estimate a model that partitions the variance of the means into within-sentence and between-sentence components. The location submodel is identical to the model in Chapter 3:

$$\log\left(\mu_i / (1 - \mu_i)\right) = \beta_0 + \beta_1 x_{1i} + \beta_2 x_{2i} + \beta_3 x_{3i} + \beta_4 x_{4i} + \beta_5 x_{5i}, \quad (6.1)$$

where the first four $x_{ji}, j = 1, 2, 3, 4$ denote the languages (as listed in Table 6.6) and x_{5i} is sentence length.

On the other hand, the dispersion submodel is

$$\phi_i = \frac{\mu_i / (1 - \mu_i)}{\sigma_i^2} - 1 = \frac{\mu_i / (1 - \mu_i)}{\sigma_{wi}^2 + \sigma_b^2} - 1, \quad (6.2)$$

where σ_{wi}^2 denotes the within-sentence variance of the ith sentence and σ_b^2 denotes the between-sentence variance, the latter being estimated by $\sigma_b^2 = \exp(\delta)$. Thus, the dispersion submodel in equation (6.2) models the precision parameter ϕ_i via a variance-partition model appropriate for the conditional beta distribution. For the sake of clarity, we have omitted language and sentence-length effects from this dispersion submodel.

The MCMC estimation procedure used a two-chains model with a 2,000-iteration burn-in and 6,000 iterations for the model estimation. Convergence was indicated by the chains mixing well, reasonably normal-distributed posterior distributions for each of the parameters, and the Gelman-Rubin $\hat{R}$ values very close to 1 for each parameter, indicating model convergence for the parameter estimates. Table 6.6 displays the resulting coefficient estimates, standard errors, and 95% credible intervals.

The average within-sentence variance is 0.0059, while the between-sentence variance estimate is $\exp(-3.181) = 0.0415$, which is greater by

Table 6.6 MCMC Model Coefficients and 95% Credible Intervals

	Coefficient	Estimate	SE	Lower	Upper
Intercept	β_0	1.647	0.184	1.285	2.006
Spanish	β_1	−0.249	0.153	−0.550	0.047
Japanese	β_2	−1.526	0.161	−1.841	−1.209
Norwegian	β_3	−0.391	0.148	−0.674	−0.104
Chinese	β_4	−1.209	0.167	−1.532	−0.877
Length	β_5	−0.287	0.103	−0.487	−0.082
Precision	δ	−3.181	0.082	−3.341	−3.023

a factor of 7.034 and underscores the substantial differences in ratings across the languages. A beta GLM with effects in the location submodel but none in the precision submodel is provided in the supplementary materials for this section. Comparisons between that model and this one reveal that the coefficients have smaller magnitudes in the Bayesian model. This probably is due to the between-sentences variance component consuming some of the variation that otherwise would be attributed to language and/or sentence length.

6.5   Roads Less Traveled and the State of the Art

At present, modeling bounded variables is an incomplete enterprise. In this concluding section, we survey some of the active areas of research and summarize the state of the art for the types of models we have covered. The survey is necessarily brief but should give readers a sense of possible extensions to those models and topics remaining to be explored and developed. We already have seen earlier in this chapter that modeling dependency structures among bounded variables is an area that needs further development. We review three additional research-active topics: scale construction and data reduction with bounded variables, the treatment of bounded variables as predictors or covariates, and imputation methods for missing data with bounded variables.

Scale construction and data reduction methods, such as factor analysis (FA) or item response (IRT) models, that take bounds into account are relatively uncommon, with one major exception. Polychoric correlations sometimes are employed in FA to deal with binary or ordinal categorical variables, and there is a sizable literature on these correlations (Drasgow, 1988). Briefly, the traditional polychoric correlation involves a latent variable model that assumes the underlying construct is normally distributed. Modern treatments have relaxed that distributional assumption, permitting a semiparametric latent variable model for polychoric correlations.

Latent variable models for continuous variables with censoring also have been developed in various multivariate analysis settings but have yet to be widely used. Brown (1989) investigated the covariance structure modeling approach to assessing reliability in the presence of censoring, proposing methods to deal with the attenuation of covariances caused by censoring. Muthén (1989) appears to have been the first to present a factor model for censored data, and Waller and Muthén (1992) proposed an estimation procedure for confirmatory factor analysis models for censored data. Kamakura and Wedel (2001) extended this work and integrated the exploratory and confirmatory FA models involved. Zhou and Liu (2009) improved the earlier estimation procedures by presenting an EM algorithm, and Costa et al. (2014) provided an exact EM algorithm and a Bayesian hierarchical version of these models.

Doubly bounded variables also have received some treatment in the development of IRT models. Early IRT models for continuous response data assume multivariate normality (Samejima, 1974). Indeed, one could transform doubly bounded data to the real line (e.g., via the logit transformation of variables in the (0, 1) interval) and employ the Samejima model in the same spirit as the Johnson logit-normal distribution. This approach has been investigated and refined by several researchers (e.g., Bejar, 1977; Ferrando, 2001; Merkle et al., 2016; Müller, 1987). However, Noel and Dauvier (2007) present a response model directly applicable to continuous bounded responses using the beta distribution. Noel (2014) followed this up by generalizing the beta distribution model to a Dirichlet distribution, to accommodate unfolding models. Given the recent expansion of distributions for modeling doubly bounded random variables (Smithson & Shou, 2017), there are prospects for more IRT models of this kind.

We now turn to the treatment of bounded variables as predictors or covariates. Despite the large volume of literature and well-developed techniques for dealing with bounded dependent variables, the literature on how to treat bounded covariates is scant (other than the aforementioned FA literature). Austin and Brunner (2003) demonstrated that OLS regression models with a predictor subject to a ceiling effect can inflate Type I error rates for other predictors. More generally, as Rigobon and Stoker (2007) point out, an OLS regression model that includes a censored predictor but ignores censoring will yield biased coefficients for all predictors.

An intuitively appealing approach is to include the censoring dummy variable in the regression to model the effect of the censored observations. However, Rigobon and Stoker (2007) show that doing so misspecifies the regression model because the coefficient of the dummy variable then covaries with the other predictors in the model. The approaches that they favor exploit some of the techniques and insights from the imputation of missing values literature. On the other hand, Austin and Hoch (2004) report that maximum likelihood estimation of a model using the full sample and OLS using only the uncensored observations yields only small biases in parameter estimates, and they recommend either of these methods for dealing with censored independent variables.

It should be noted that this literature is small but nevertheless not well connected. Rigobon and Stoker do not mention Austin's papers, and a recent study of such models (Lima Taga & Singer, 2018) cites Austin and Hoch but not Rigobon and Stoker. Moreover, several important aspects of this topic remain neglected. For instance, the behavior of moderator effects involving censored predictors seems to have been unexamined thus far. There is no space here to elucidate and compare these alternative approaches, but we encourage readers to familiarize themselves with the relevant literature on this topic.

The last topic is multiple imputation. As Geraci and McLain (2018) observe, multiple imputation methods for continuous variables generally treat the variables as if they are normally distributed, thereby ignoring scale bounds. This assumption renders conventional imputation methods vulnerable to misspecification and out-of-range imputations (e.g., negative reaction times, or percentages that exceed 100). It also ignores the nature of the bounds themselves (e.g., whether cases on the boundary are censored or true scores).

There is a small and scattered literature on imputation for bounded variables. Buuren and Groothuis-Oudshoorn (2010) recommend replacing out-of-range imputations with the nearest bound (i.e., censoring the imputed value), which is the obverse of procedures for modeling censored data via standard imputation methods (Pan, 2000). Rodwell et al. (2014) demonstrates that this approach can bias estimates, especially when the data are skewed. A second approach is to employ a truncated normal regression model (Schafer, 1997). An obvious limitation to this approach is its assumption of normality, especially for strongly skewed data. Some authors have recommended transforming bounded continuous variables to the real line (Lee & Carlin, 2010) and imputing values for the transformed variable. A drawback to this tactic is that outlying values close to a boundary may have extreme negative or positive values when transformed (Hippel, 2013), and such transformations may be undefined for values on the boundary. Predictive mean matching (PMM) has been found to be useful for bounded data (Lee & Carlin, 2017). However, PPM does not make imputations beyond the *observed* range of the data (vs. the theoretical scale range).

Imputation via alternative regression models using distributions with bounded supports has not been entirely neglected in this literature, although it has appeared later than the proposals mentioned above. Several authors have proposed imputation to either singly or doubly bounded variables using appropriate distributions such as Weibull or beta distributions (Demirtas & Hedeker, 2008). The primary drawback of these models, like the conventional normal-theory regression model, is that they assume the data follow a particular distribution. Geraci and McLain (2018) propose a semiparametric imputation model using quantile regression as its basis, combined with a fitted transformation that ensures within-bounds imputed values. Its main practical disadvantage is its computational intensity. Geraci and McLain's simulation study suggests that their approach performs well by comparison with some other methods, even with PPM, although for small samples, its advantages are unclear.

Of all the imputation methods for bounded variables described here, PPM, parametric regression models using distributions with bounded support, and the Geraci-McLain semiparametric approach would seem to be the only ones worth serious consideration. Nonetheless, it should be borne in mind that, as of this writing, imputation for bounded variables still is in an early stage of development, and it appears unlikely that any single method will dominate the others under all conditions. Issues that remain to be systematically considered here include the nature of the bounds (e.g., censored vs. true score) and boundary inflation.

6.5.1 State of the Art

A primary motivation for this book is that most research methods handbooks neglect bounded variables or treat them as exceptions or "pathological" cases. Consequently, while there is an obvious need for everyday researchers to be able to effectively analyze bounded variables, the relevant techniques are scattered across several literatures and buried in specialist journals. We have attempted to synthesize and integrate these disparate sources without overwhelming readers with technical detail or specialist jargon. We began with the most popular bounded-variable models, namely, the Tobit and lognormal models. From there we provided chapters on singly bounded and doubly bounded dependent variables and censored variables. Boundary-inflated variables were incorporated into the material on doubly bounded variables. Throughout the book, we limited coverage to bounded continuous distributions because these are typically the most neglected and misunderstood. These final paragraphs are our reflections on where these matters stand at the time of writing this book.

Understandably, the state of the art in this topic is at its most mature for analyzing a singly bounded dependent variable. Techniques and understanding are most advanced for dealing with censoring and boundary inflation, perhaps less so for absolute bounds and particularly for doubly bounded dependent variables. That said, research on doubly bounded dependent variables has seen rapid development since 2004. Applied researchers still lack clear guidance from the methodological literature on fundamental considerations, such as when to treat bounds as censored or absolute, and the distinction between a hurdle model and a boundary inflation model. We have tried to provide guidance on both issues.

Approaches to modeling multivariate bounded variables have seen substantial development since the 1990s, although techniques such as copula models still are relatively unknown among researchers in many domains. Models for dependent censoring and random-effects censoring, boundary inflation, and boundary location still are largely in developmental stages. Data reduction and scale construction with bounded variables also have progressed substantially, especially in factor-analytic models under censoring. Increased efforts have been devoted to developing IRT models for bounded continuous-response items, perhaps motivated partly by the ease with which sliders, cursor movements, and other continuous-response formats can be presented to participants online and growing recognition of the advantages of continuous-response over categorical-response formats. However, relatively little has been done on the topic of bounded covariates, and most of the developments there have focused on main-effects models. We have yet to see any full-fledged treatments of moderator effects that take censoring, boundary inflation, or absolute bounds into account.

What are the prospects for teaching about bounded variables in methods courses? As more textbooks or handbook materials on this topic become available, those prospects will improve. Likewise, although software resources take time to develop, the open-source software movement has considerably sped up developments, and the ready availability of software resources will also speed learning and adoption of techniques. Far from being exceptions or pathological cases, bounded variables actually are the norm, and they are ubiquitous throughout the human sciences. They merit a central place in the education of future researchers and research consumers, and that has been a key motivation behind this book.

REFERENCES

Agresti, A. (2010). *Analysis of ordinal categorical data* (Vol. 656). New York, NY: John Wiley.

Agresti, A. (2013). *Categorical data analysis*. New York, NY: John Wiley.

Aitchison, J. (1955). On the distribution of a positive random variable having a discrete probability mass at the origin. *Journal of the American Statistical Association, 50*(271), 901–908.

Aitkin, M. (1987). Modelling variance heterogeneity in normal regression using glim. *Applied statistics, 36*(3), 332–339.

Akaike, H. (1974). A new look at the statistical model identification. *IEEE Transactions on Automatic Control, 19*(6), 716–723.

Amemiya, T. (1984). Tobit models: A survey. *Journal of Econometrics, 24*(1–2), 3–61.

Anscombe, F. J. (1953). Contribution to the discussion of H. Hotelling's paper. *Journal of the Royal Statistical Society B, 15*, 229–230.

Aribarg, A., Pieters, R., & Wedel, M. (2010). Raising the BAR: Bias adjustment of recognition tests in advertising. *Journal of Marketing Research, 47*(3), 387–400.

Austin, P. C., & Brunner, L. J. (2003). Type I error inflation in the presence of a ceiling effect. *The American Statistician, 57*(2), 97–104.

Austin, P. C., & Hoch, J. S. (2004). Estimating linear regression models in the presence of a censored independent variable. *Statistics in Medicine, 23*(3), 411–429.

Bascoul-Mollevi, C., Gourgou-Bourgade, S., & Kramar, A. (2005). Two-part statistics with paired data. *Statistics in Medicine, 24*(9), 1435–1448.

Beck, A. T., Steer, R. A., & Brown, G. K. (1996). Beck Depression Inventory–II. *San Antonio, 78*(2), 490–498.

Bera, A. K., Jarque, C. M., & Lee, L. F. (1984). Testing the normality assumption in limited dependent variable models. *International Economic Review, 25*, 563–578.

Breen, R. (1996). *Regression models: Censored, sample selected, or truncated data*. Thousand Oaks, CA: Sage.

Brown, R. (1989). Congeneric modeling of reliability using censored variables. *Applied Psychological Measurement, 13*(2), 151–159.

Budescu, D. V., Por, H.-H., Broomell, S. B., & Smithson, M. (2014). The interpretation of IPCC probabilistic statements around the world. *Nature Climate Change, 4*(6), 508.

Buuren, S. v., & Groothuis-Oudshoorn, K. (2010). mice: Multivariate imputation by chained equations in R. *Journal of Statistical Software, 45*(3), 1–68.

Chesher, A., & Irish, M. (1987). Residual analysis in the grouped and censored normal linear model. *Journal of Econometrics, 34*, 33–61.

Collet, D. (2003). *Modelling binary data*. Boca Raton, FL: Chapman & Hall.

Cook, R. D. (1977). Detection of influential observations in linear regression. *Technometrics, 19*, 15–18.

Cordeiro, G. M., Da Rocha, E. C., Da Rocha, J. G. C., & Cribari-Neto, F. (1997). Bias-corrected maximum likelihood estimation for the beta distribution. *Journal of Statistical Computation and Simulation, 58*(1), 21–35.

Costa, D. R., Lachos, V. H., Bazan, J., & Azevedo, C. L. (2014). Estimation methods for multivariate Tobit confirmatory factor analysis. *Computational Statistics & Data Analysis, 79*, 248–260.

Coste, C., Meunier, F., Novik, N., Reeves, M., & Tjong, E. (2018). Starting a business: Transparency of information at business registries. In *Doing business 2018: Reforming to create jobs*. Washington, DC: The World Bank.

Cox, D. R., & Snell, E. J. (1989). *Analysis of binary data*. New York, NY: Chapman and Hall.

Cribari-Neto, F., & Zeileis, A. (2010). Beta regression in R. *Journal of Statistical Software, 34*(2), 1–24. Available from http://www.jstatsoft.org/v34/i02/

Daoud, M. (2007). *Extensions of two-part tests to compare k independent populations*. Western Michigan University. Retrieved from https://scholarworks.wmich.edu/dissertations/848.

Demirtas, H., & Hedeker, D. (2008). Imputing continuous data under some non-gaussian distributions. *Statistica Neerlandica, 62*(2), 193–205.

Drasgow, F. (1988). Polychoric and polyserial correlations. *Encyclopedia of statistical sciences*.

Engle, R. F. (1984). Wald, likelihood ratio, and Lagrange multiplier tests in econometrics. In *Z. Griliches & M. D. Intriligator (Eds.), Handbook of econometrics* (Vol. II). Elsevier.

Epsinheira, P., Ferrari, S., & Cribari-Neto, F. (2008). On beta regression residuals. *Journal of Applied Statistics, 35*, 407–419.

Everett, J. A. C. (2013). The 12 item social and economic conservatism scale (SECS). *PLoS One, 8*(12), e82131.

Ferrari, S. L. P., & Cribari-Neto, F. (2004). Beta regression for modeling rates and proportions. *Journal of Applied Statistics, 10*, 1–18.

Firth, D. (1993). Bias reduction of maximum likelihood estimates. *Biometrika, 80*, 27–38.

Geraci, M., & McLain, A. (2018). Multiple imputation for bounded variables. *Psychometrika, 83*(4), 1–22.

Gill, J. (2000). *Generalized linear models: A unified approach* (Vol. 134). Thousand Oaks, CA: Sage.

Gupta, A. K., & Nadarajah, S. (2004). *Handbook of beta distribution and its applications*. Boca Raton, FL: CRC Press.

Hallinan, A. J. (1993). A review of the Weibull distribution. *Journal of Quality Technology, 25*(2), 85–93.

Hao, L., & Naiman, D. Q. (2007). *Quantile regression* (No. 149). Thousand Oaks, CA: Sage.

Harris, P. S., Harris, P. R., & Miles, E. (2017). Self-affirmation improves performance on tasks related to executive functioning. *Journal of Experimental Social Psychology, 70*, 281–285.

Hippel, P. T. von. (2013). Should a normal imputation model be modified to impute skewed variables? *Sociological Methods & Research, 42*(1), 105–138.

Holden, D. (2004). Testing the normality assumption in the Tobit model. *Journal of Applied Statistics, 31*(5), 521–532.

Holden, D. (2011). Testing for heteroskedasticity in the Tobit and probit models. *Journal of Applied Statistics, 38*(4), 735–744.

Hosmer Jr., D. W., Lemeshow, S., & Sturdivant, R. X. (2013). *Applied logistic regression* (Vol. 398). New York, NY: John Wiley.

Huang, C. J., Sloan, F. A., & Adamache, K. W. (1987). Estimation of seemingly unrelated tobit regressions via the em algorithm. *Journal of Business & Economic Statistics, 5*(3), 425–430.

Jarque, C. (1981). A test for heteroscedasticity in a limited dependent variable model. *Australian Journal of Statistics, 23*(2), 159–163.

Johnson, N., Kotz, S., & Balakrishnan, N. (1994). *Continuous univariate probability distributions* (Vol. 1). New York, NY: John Wiley.

Johnson, N. L., Kotz, S., & Balakrishnan, N. (1995). *Continuous univariate distributions* (Vol. 2). New York, NY: John Wiley.

Kamakura, W. A., & Wedel, M. (2001). Exploratory tobit factor analysis for multivariate censored data. *Multivariate Behavioral Research, 36*(1), 53–82.

Koenker, R., & Bassett Jr., G. (1978). Regression quantiles. *Econometrica: Journal of the Econometric Society, 46*(1), 33–50.

Kruschke, J. (2014). *Doing Bayesian data analysis: A tutorial with R, JAGS, and Stan*. New York, NY: Academic Press.

Lambert, D. (1992). Zero-inflated Poisson regression, with an application to defects in manufacturing. *Technometrics, 34*(1), 1–14.

Lau, J. H., Clark, A., & Lappin, S. (2014). Measuring gradience in speaker grammaticality judgements. In *Proceedings of the annual meeting of the cognitive science society* (Vol. 36). Quebec City, Canada: Cognitive Science Society.

Lee, K. J., & Carlin, J. B. (2010). Multiple imputation for missing data: fully conditional specification versus multivariate normal imputation. *American journal of Epidemiology, 171*(5), 624–632.

Lee, K. J., & Carlin, J. B. (2017). Multiple imputation in the presence of non-normal data. *Statistics in Medicine, 36*(4), 606–617.

Lima Taga, M. F. de, & Singer, J. M. (2018). Simple linear regression with interval censored dependent and independent variables. *Statistical Methods in Medical Research, 27*(1), 198–207.

Long, J. S. (1997). *Regression models for categorical and limited dependent variables*. Thousand Oaks, CA: Sage.

Marshall, A. W., & Olkin, I. (2007). *Life distributions* (Vol. 13). New York, NY: Springer.

McCullagh, P., & Nelder, J. A. (1989). *Generalized linear models* (2nd ed.). Boca Raton, FL: Chapman and Hall/CRC.

McElreath, R. (2015). *Statistical rethinking*. Boca Raton, FL: CRC Press.

Menard, S. (2000). Coefficients of determination for multiple logistic regression analysis. *The American Statistician, 54*(1), 17–24.

Mullahy, J. (1986). Specification and testing of some modified count data models. *Journal of Econometrics, 33*(3), 341–365.

Muthén, B. O. (1989). Tobit factor analysis. *British Journal of Mathematical and Statistical Psychology, 42*(2), 241–250.

Nelsen, R. B. (1999). *An introduction to copulas, volume 139 of lecture notes in statistics*. New York, NY: Springer-Verlag.

Noel, Y. (2014). A beta unfolding model for continuous bounded responses. *Psychometrika, 79*(4), 647–674.

Noel, Y., & Dauvier, B. (2007). A beta item response model for continuous bounded responses. *Applied Psychological Measurement, 31*(1), 47–73.

Ospina, R., Cribari-Neto, F., & Vasconcellos, K. L. (2006). Improved point and interval estimation for a beta regression model. *Computational Statistics and Data Analysis, 51*, 960–981.

Ospina, R., & Ferrari, S. L. P. (2012). A general class of zero-or-one inflated beta regression models. *Computational Statistics and Data Analysis, 56*, 1609–1623.

Pan, W. (2000). A multiple imputation approach to cox regression with interval-censored data. *Biometrics, 56*(1), 199–203.

Paolino, P. (2001). Maximum likelihood estimation of models with beta-distributed dependent variables. *Political Analysis, 9*(4), 325–346.

Pimentel, R. S., Niewiadomska-Bugaj, M., & Wang, J.-C. (2015). Association of zero-inflated continuous variables. *Statistics & Probability Letters, 96*, 61–67.

Powell, J. (1984). Least absolute deviations estimation for the censored regression model. *Journal of Econometrics, 25*(3), 303–325.

Rabe-Hesketh, S., Yang, S., & Pickles, A. (2001). Multilevel models for censored and latent responses. *Statistical Methods in Medical Research, 10*(6), 409–427.

Rigby, R. A., & Stasinopoulos, D. M. (2005). Generalized additive models for location, scale and shape (with discussion). *Applied Statistics, 54*, 507–554.

Rigobon, R., & Stoker, T. M. (2007). Estimation with censored regressors: Basic issues. *International Economic Review, 48*(4), 1441–1467.

Rodwell, L., Lee, K. J., Romaniuk, H., & Carlin, J. B. (2014). Comparison of methods for imputing limited-range variables: A simulation study. *BMC medical research methodology, 14*(1), 57.

Samejima, F. (1974). Normal ogive model on the continuous response level in the multidimensional latent space. *Psychometrika, 39*(1), 111–121.

Schafer, J. L. (1997). *Analysis of incomplete multivariate data.* Chapman and Hall/CRC.

Schwarz, G. (1978). Estimating the dimension of a model. *The Annals of Statistics, 6*(2), 461–464.

Shou, Y., & Smithson, M. (2019). cdfquantreg: An R package for CDF-quantile regression. *Journal of Statistical Software, 88*(1), 1–30.

Shou, Y., & Song, F. (2017). Decisions in moral dilemmas: The influence of subjective beliefs in outcome probabilities. *Judgment & Decision Making, 12*(5).

Simas, A. B., Barreto-Souza, W., & Rocha, A. V. (2010). Improved estimators for a general class of beta regression models. *Computational Statistics and Data Analysis, 54*, 348–366.

Sklar, M. (1959). Fonctions de repartition a n dimensions et leurs marges. *Publ. Inst. Statist. Univ. Paris, 8*, 229–231.

Smithson, M., & Merkle, E. C. (2013). *Generalized linear models for categorical and continuous limited dependent variables.* Boca Raton, FL: CRC Press.

Smithson, M., & Shou, Y. (2014). Randomly stopped sums: Models and psychological applications. *Frontiers in Psychology, 5.*

Smithson, M., & Shou, Y. (2017). CDF-quantile distributions for modelling random variables on the unit interval. *British Journal of Mathematical and Statistical Psychology, 70*(3), 412–438.

Smithson, M., & Verkuilen, J. (2006). A better lemon squeezer? Maximum likelihood regression with beta-distributed dependent variables. *Psychological Methods, 11*, 54–71.

Stasinopoulos, D. M., & Rigby, R. A. (2007). Generalized additive models for location scale and shape (GAMLSS) in r. *Journal of Statistical Software, 23*(7), 1–46.

Tadikamalla, P., & Johnson, N. L. (1982). Systems of frequency curves generated by transformations of logistic variables. *Biometrika, 69*(2), 461–465.

Tobin, J. (1958). Estimation of relationships for limited dependent variables. *Econometrica: Journal of the Econometric Society*, 24–36.

Trivedi, P. K., & Zimmer, D. M. (2007). Copula modeling: An introduction for practitioners. *Foundations and Trends® in Econometrics, 1*(1), 1–111.

Tweedie, M. (1984). An index which distinguishes between some important exponential families. In *Statistics: Applications and new directions: Proc. indian statistical institute golden jubilee international conference* (Vol. 579, p. 604). Calcutta: Indian Statistical Institute.

U.S. Census Bureau. (2015). *American Community Survey, Five-Year Public Use Microdata Sample, 2010–2015.* Retrieved from https://www.census.gov/programs-surveys/acs

Verkuilen, J., & Smithson, M. (2012). Mixed and mixture regression models for continuous bounded responses using the beta distribution. *Journal of Educational and Behavioral Statistics, 37*(1), 82–113.

Vrieze, S. I. (2012). Model selection and psychological theory: A discussion of the differences between the Akaike information criterion (AIC) and the Bayesian information criterion (BIC). *Psychological Methods, 17*(2), 228.

Waller, N. G., & Muthén, B. O. (1992). Genetic tobit factor analysis: Quantitative genetic modeling with censored data. *Behavior Genetics, 22*(3), 265–292.

Weber, E. U., Blais, A.-R., & Betz, N. E. (2002). A domain-specific risk-attitude scale: Measuring risk perceptions and risk behaviors. *Journal of Behavioral Decision Making, 15*(4), 263–290.

Wong, K. Y., Zeng, D., & Lin, D. (2017). Efficient estimation for semiparametric structural equation models with censored data. *Journal of the American Statistical Association, 113*(522), 893–905.

World Values Survey Association. (2015). *World Values Survey, Wave 6, 2010–2014.* Retrieved from https://www.worldvaluessurvey.org

Yu, Y., Yang, X., Yang, Y., Chen, L., Qiu, X., Qiao, Z., et al. (2015). The role of family environment in depressive symptoms among university students: A large sample survey in China. *PLoS One, 10*(12), e0143612.

Zhou, X., & Liu, X. (2009). The Monte Carlo EM method for estimating multivariate Tobit latent variable models. *Journal of Statistical Computation and Simulation, 79*(9), 1095–1107.

INDEX